Experiment, right or wrong

Experiment, right or wrong

Allan Franklin
Department of Physics
University of Colorado

The right of the
University of Cambridge
to print and sell
all manner of books
was granted by
Henry VIII in 1534.
The University has printed
and published continuously
since 1584.

Cambridge University Press

Cambridge
New York   Port Chester   Melbourne   Sydney

CAMBRIDGE UNIVERSITY PRESS
Cambridge, New York, Melbourne, Madrid, Cape Town, Singapore, São Paulo

Cambridge University Press
The Edinburgh Building, Cambridge CB2 8RU, UK

Published in the United States of America by Cambridge University Press, New York

www.cambridge.org
Information on this title: www.cambridge.org/9780521382076

First published 1990
This digitally printed version 2008

A catalogue record for this publication is available from the British Library

Library of Congress Cataloguing in Publication data
Franklin, Allan.
Experiment, right or wrong / Allan Franklin.
p. cm.
ISBN 0-521-38207-6
1. Science – Philosophy. 2. Science – History. 3. Nuclear physics –
Experiments. I. Title.
Q175.F785 1990
501 – dc20   89-23848
                                                                    CIP

ISBN 978-0-521-38207-6 hardback
ISBN 978-0-521-06477-4 paperback

Contents

Preface

In an interview filmed just a few weeks before his death, Richard Feynman described his search for Tannu Tuva. Tannu Tuva was a small country in Asia, now part of the Soviet Union, known in the West primarily for its triangular postage stamps. Feynman wanted to visit Tannu Tuva, but didn't want to take advantage of his reputation as a Nobel Prize–winning physicist. He wanted to do it, as he put it, "in the right way." His efforts to arrange a visit spanned several years. In the process he learned to write Tuvan and became familiar with Tuvan singing. Feynman's invitation to Tannu Tuva arrived two weeks after his death.

Science is like Feynman's quest. The most important thing is to do it "in the right way."

Acknowledgments

In my last book I noted my debts to a large number of people who had helped me in changing my area of research from physics to the history and philosophy of science. Those debts are still unpaid. There are several people who have, in the past three years, increased my indebtedness.

First and foremost is my friend and research collaborator, Colin Howson. Our continuing work together has taught me much of what I know about the Bayesian approach to the philosophy of science, and his comments and constructive criticism have been invaluable. His friendship and support have been even more important.

I am particularly grateful to Robert Ackermann for having read this manuscript and for his constructive comments and helpful discussions. Any errors that remain are, of course, my own responsibility. There are several other scholars who have also taken the time to give my work a sympathetic reading and to offer valuable criticism. These include Peter Galison, Ian Hacking, Howard Smokler, Jim Cushing, Andy Pickering, Heinz Post, and Michael Redhead.

I am grateful to my colleague Brian Ridley, a major participant in the angular correlation experiments during the 1950s, for many valuable discussions, comments, and criticisms, and also for providing me with his files from that period, which included the unpublished paper of Wu and Schwarzschild. Robert Marshak provided a copy of his unpublished paper on the history of the *V-A* theory and helpful conversations. Stanley Ruby also provided extremely valuable insights into the history of the angular correlation experiments.

Some of this work was done while I was a research fellow at

the Center for Philosophy of Science, University of Pittsburgh. I am grateful to the Center for its support and hospitality.

Portions of this book were originally published elsewhere. Chapter 1 appeared in "Experiment and the Development of the Theory of Weak Interactions: Fermi's Theory," *PSA 1986/2*: 163–79. Part of Chapter 6 appeared in "It Probably Is a Valid Experimental Result: A Bayesian Approach to the Epistemology of Experiment," with Colin Howson, *Studies in History and Philosophy of Science* 19 (1988): 419–27. I thank the Philosophy of Science Association and Pergamon Press, which hold the copyrights, for permission to reproduce the material in this book.

Introduction

Scientists need no convincing that experiment plays an essential role in science. It provides the basis for theory choice, confirms or refutes hypotheses or theories, and sometimes calls for new theories. These are only a few of its roles and, as Ian Hacking (1983) has pointed out, experiment often has a life of its own. Nevertheless, I called my previous book *The Neglect of Experiment* (1986). Who was neglecting experiment? Certainly not scientists. I believed then that it was historians, philosophers, and sociologists of science. Even among those who acknowledged the importance of experimental results there tended to be an almost mythological treatment of a few standard exemplary experiments, such as Galileo and the Leaning Tower, Young's double slit interference experiment, and the Michelson-Morley experiment. Actual experiments were rarely discussed.

Fortunately, this is no longer the case. One of the most interesting and exciting trends in history, philosophy, and sociology of science in the 1980s has been the study of and emphasis on actual experiments. Philosophers such as Dudley Shapere (1982), Ian Hacking (1983), Nancy Cartwright (1983), and Robert Ackermann (1985) have used the actual practice of science to analyze and illuminate what good science should be. Historians such as Bruce Wheaton (1983), Peter Galison (1987), and Roger Stuewer (1975) have not only provided us with detailed studies of particular experiments, but have also given us new perspectives on the role of real experiments. Sociologists of science have added to our knowledge by their detailed studies of experiments. Although, as discussed in detail later, I disagree with their view that science is merely a social construction, there is no doubt that the work of Harry Collins

(1976), Andy Pickering (1984a, 1984b), and Trevor Pinch (1986) has enhanced our knowledge of experiment.[1]

Hacking began the discussion of how we come to believe rationally, or reasonably, in experimental results. Cartwright's emphasis on the actual practice of science has given us new insights into the nature of scientific explanation. Galison emphasizes the continuity of scientific instruments and of experimental apparatus and practice. He notes that changes in these rarely occur at the same time as major theoretical changes. Thus, they provide a continuous empirical basis for science, one that the usual theory-dominated accounts fail to give. Ackermann has also discussed the role of instruments in providing stable data.

In this book I will argue that the practice of science is reasonable.[2] This has been implicit in my previous work and I will discuss it explicitly here, and contrast it to the view that science is merely a social construction. The view of science I propose is what one might call an "evidence model" of science. I suggest that when questions of theory choice, confirmation, or refutation are raised they are answered on the basis of valid experimental evidence. I will also argue that there are good reasons for belief in the validity of that evidence. This is both a descriptive and a normative view. I believe that the history of science, as illustrated in the episodes presented here and in my (1986) book, demonstrates that scientists behave this way. I also believe that this is the way they should behave. The Bayesian approach to the philosophy of science, which I believe is both a fruitful way of looking at science and also part of a theory of rationality, requires that observation of evidence entailed by a hypothesis strengthen our belief in that hypothesis. Although I know of no episodes in the history of science in which scientific decisions have gone against the weight of evidence,[3] I think that scientists in that case would have been unreasonable.[4]

[1] I have restricted myself here to discussions of twentieth-century experiments. There have also been several conferences and books devoted to the study of theory and experiment. See Achinstein and Hannaway (1985), Batens and van Bendegem (1988), and Gooding, Pinch, and Schaffer (1988).
[2] I will use the term "reasonable" rather than "rational" because I do not have a complete theory of rationality. I do believe, however, that part of such a theory includes the idea that observation of evidence entailed by a hypothesis should strengthen our belief in the hypothesis.
[3] I am referring here to the context of justification and not to the context of

There will, of course, be cases in which the evidence available at a given time may not be sufficient to decide an issue, or in which there are arguments as to what constitutes relevant evidence or on the validity of the evidence, but this is to be expected. Such episodes themselves assume an evidence model. Even in cases of scientific fraud, those who engage in the fraud are behaving in the way suggested by an evidence model by forging, cooking, or trimming their results to support their views. (See Franklin 1986, ch. 8, for details of some episodes of scientific fraud.)

I will continue my study of two questions. (1) What role does and should experiment play in the choice between competing theories and in the confirmation or refutation of theories and hypotheses? (2) How do we come to believe reasonably in experimental results? I hope to provide a fuller and more detailed philosophy of experiment than I have previously, including a somewhat different approach to the Duhem-Quine problem, or the localization of support or refutation. I will show that my previously presented epistemology of experiment, a set of strategies for reasonable belief in experimental results, can be explicated in terms of Bayesian confirmation theory. I will further argue that this, as well as other evidence, makes Bayesianism a fruitful way to look at science. I will also provide additional historical case studies, because I believe that the philosophy of science benefits from the study of actual science. One major difference will be the consideration of the fallibility and corrigibility of experimental results. In my previous case studies I dealt with episodes in which the experimental results are still regarded as valid and correct by the physics community. For example, subsequent work has not cast doubt on the experiments that demonstrated the nonconservation of parity or *CP* violation.[5] (See Franklin 1986, chs. 1 and 3, for details.)

discovery or to what one might call the context of pursuit, the decision of a scientist to pursue a certain program of research.

[4] Bayesianism requires that you change your degree of belief using conditionalization. If you don't conditionalize there are bets that can be made against you such that you will always lose money.

[5] Even though the average value of the *CP*-violating parameter, η_{+-}, has changed from $(1.95 \pm 0.03) \times 10^{-3}$ to $(2.27 \pm 0.022) \times 10^{-3}$ the existence of *CP* violation has not been questioned. This change in the average value of this quantity is also an example of experimental fallibility. See Chapter 6 for

Episodes such as these often occur in the history of physics, giving us confidence in the strategies used to validate experimental results. Often, however, the history of physics shows that experimental results are both fallible and corrigible. In dealing with the complex interaction between theory and experiment the old cliché "Man proposes, Nature disposes" has been shown to be far from adequate. John Worrall's (1982) account of the nineteenth-century experiments that attempted to measure the pressure of light shows that it is not only difficult to know what man is proposing, but that it is also difficult to learn what Nature's disposition is. Not only did the experimental results change, but also their theoretical interpretation. The studies of the experiments that demonstrated the existence of weak neutral currents, by Galison (1987, ch. 4) and by Pickering (1984b), illustrate the same point. During the 1960s, events were seen that, in the light of later theoretical developments and experimental studies, are now interpreted as providing evidence for the existence of weak neutral currents. At the time, however, they were thought to be caused by neutron background. During the 1930s, experimental results were reported that, in retrospect, illustrate nonconservation of parity (Franklin 1986, ch. 2.) Their significance was not realized by the experimenters or by anyone else for more than twenty-five years. It was only after the discovery of parity nonconservation in the 1950s that these results were reinterpreted. In addition, history shows that the experimental results were first thought to be valid, then believed to be an instrumental artifact, then again thought to be correct, then believed again to be an artifact, and are currently thought to be correct. The interpretation and validation of experimental results is not a simple task.

In this study I will present detailed histories of two such episodes: (1) the interaction of experiment and theory in the development of the theory of weak interactions from Fermi's theory in 1934 to the *V-A* theory of 1957 and (2) atomic parity violating experiments in the 1970s and 1980s and their interaction with the Weinberg-Salam unified theory of electroweak interactions. In these episodes we will see not only that experimental results can

details. The original result of Christenson et al. (1964) of $(2.0 \pm 0.4) \times 10^{-3}$ is consistent with either average value.

be wrong, but also that theoretical calculations and the comparison between experiment and theory can also be incorrect.

Does the fallibility and corrigibility of experiment, of theory, and of their comparison affect our answers to the two questions posed earlier? Can we still maintain that experiment plays a legitimate role in theory choice and confirmation? Can we still argue that there are good strategies for reasonable belief in experimental results? I believe the answer to all of these questions is yes, and my arguments follow.

I

EXPERIMENT AND THE DEVELOPMENT OF THE THEORY OF WEAK INTERACTIONS: FROM FERMI TO *V-A*

1

Fermi's theory

The fallibility and corrigibility of experimental results, of theoretical calculation, and of the comparison between experiment and theory will be amply illustrated in the episode to be discussed. This section will deal with the relation between experiment and theory in the field of weak interactions during the period between Fermi's proposal of his theory of β decay in 1934 and the acceptance of the V-A theory of weak interactions in 1959. Part of the fascination of this story is that the V-A theory appeared to be refuted by existing experimental evidence at the time it was proposed by Sudarshan and Marshak (1957) and by Feynman and Gell-Mann (1958). The authors, themselves, recognized this and suggested that the experimental results might be wrong, a suggestion that turned out to be correct. Nevertheless the theory was proposed because it seemed to be the only available candidate for a universal theory of the weak interaction. In this section I will examine the origin and development of this idea of a universal theory of the weak interaction to the acceptance of the V-A theory as such a theory.

Fermi's (1934a, 1934b) theory of β decay was introduced in 1934. It was not the first quantitative theory of β decay. Beck and Sitte (1933) had formulated an earlier theory using Dirac's prediction of the positron. According to their 1933 model an electron–positron pair was created. The positron was absorbed by the nucleus and the electron emitted (or vice versa). This theory had currency for a short time, but was rejected on experimental grounds, to be discussed below. Fermi assumed the existence of the neutrino, then recently proposed by Pauli,[1] and

---

[1] Pauli had originally called the particle the neutron, but following Chadwick's discovery of a heavy neutral particle, Fermi coined the name neutrino, or little neutral one.

used the method of second quantization. He also assumed, with Heisenberg, that the nucleus contained only protons and neutrons and that the electron and the neutrino were created at the instant of decay. This was because no theory available at that time[2] could explain how an electron and a neutrino could be bound inside the nucleus. Fermi added a perturbing energy due to the decay interaction to the Hamiltonian of the nuclear system. In modern notation this perturbation is of the form

$$H_{if} = G \left[U^*_f \phi_e(r) \phi_\nu(r) \right] O_x U_i, \tag{1.1}$$

where U_i and U_f describe the initial and final states of the nucleus, ϕ_e and ϕ_ν are the electron and antineutrino wavefunctions, respectively, and O_x is a mathematical operator.

Pauli (1933) had previously shown that O_x can take on only five forms if the Hamiltonian is to be relativistically invariant. We identify these as S, the scalar interaction; P, pseudoscalar; V, polar vector; A, axial vector; and T, tensor.[3] Fermi knew this, but, in analogy with electromagnetic theory, and because his calculations were in agreement with experiment, he chose to use only the vector form of the interaction. He also considered only what he called "allowed" transitions, those for which the electron and neutrino wavefunctions could be considered constant over nuclear dimensions. He recognized that "forbidden" transitions

[2] There are some contemporary theories that do allow an electron and a neutrino to be bound inside a nucleus. See, for example, Barut (1980, 1982). These theories are not widely accepted within the physics community.

[3] We wish to consider the relativistically invariant combinations of the wave functions. Let U_f and U_i represent the initial and final nuclear states and ϕ_e and ϕ_ν be the electron and antineutrino wavefunctions, respectively. Let Q be an operator which, when applied to the wavefunction describing the initial nuclear state, substitutes for it one in which a proton replaces a neutron. Q^* causes the nucleon to make the opposite transition. The five allowable interactions are:

Scalar: $S = (U^*_f \beta Q_k U_i)(\phi^*_e \beta \phi_\nu)$,
Vector: $V = (U^*_f Q_k U_i)(\phi^*_e \phi_\nu) - (U^*_f \alpha Q_k U_i)(\phi^*_e \alpha \phi_\nu)$,
Tensor: $T = (U^*_f \beta \sigma Q_k U_i)(\phi^*_e \beta \sigma \phi_\nu) + (U^*_f \beta \alpha Q_k U_i)(\phi^*_e \beta \alpha \phi_\nu)$,
Axial Vector: $A = (U^*_f \sigma Q_k U_i)(\phi^*_e \sigma \phi_\nu) - (U^*_f \gamma_5 Q_k U_i)(\phi^*_e \gamma_5 \phi_\nu)$,
Pseudoscalar: $P = (U^*_f \gamma_5 Q_k U_i)(\phi^*_e \beta \gamma_5 \phi \nu)$.

α is a vector whose three components are the Dirac matrices. σ differs from the usual Pauli spin matrices only in being doubled to four rows and four columns. β is the fourth Dirac matrix, and $\gamma_5 = -i\alpha_x \alpha_y \alpha_z$. See Konopinski (1943) for details.

would also exist. The rate of such transitions would be much reduced and the shape of the spectrum would differ from that of the allowed transitions. He found that for allowed transitions certain selection rules would apply. These included: no change in the angular momentum of the nucleus, $\Delta J = 0$, and no change in the parity (space reflection properties) of the nuclear states. He also found for such transitions, assuming the mass of the neutrino was zero, that

$$P(W)\, dW = G^2 \, |M|^2\, f(Z,W)\, (W_o - W)^2\, (W^2 - 1)^{1/2}\, W\, dW,$$
$$(1.2)$$

where W is the energy of the electron (in units of $m_e c^2$), W_o is the maximum energy allowed, $P(W)$ is the probability of the emission of an electron with energy W, and $f(Z,W)$ is a function giving the effect of the Coulomb field of the nucleus on the emission of electrons. It was later shown that for allowed transitions the energy dependence of the β spectrum was independent of the choice of interaction (Konopinski 1943). Fermi also showed that the value of $F(Z,W_o)\tau_o$ should be approximately constant for each type of transition, that is, allowed, first forbidden, second forbidden, and so forth. $F(Z,W_o)$ is the integral of the energy distribution and τ_o is the lifetime of the transition. Fermi cited already published experimental results in support of his theory, in particular the work of Sargent on the shape of β-decay spectra (Sargent 1932) and on decay constants and maximum electron energies (Sargent 1933). Sargent had found that if he plotted the logarithm of the disintegration constants (inversely proportional to the lifetime) against the logarithm of the maximum electron energy, the data for all measured decays fell into two distinct groups, known in the later literature as Sargent curves (Figure 1.1). Although Sargent had remarked that, "At present the significance of this general relation is not apparent" (1933, p. 671), that was what Fermi's theory required, namely, that $F\tau_o$ is approximately constant for each type of decay. (Note that the value of F depends on W_o, the maximum electron energy.) Fermi associated the two Sargent curves with the allowed and first forbidden transitions, in analogy with electromagnetic dipole and quadrupole radiation. The general shape of the observed spectra also agreed with Fermi's model.

Although Fermi's theory had received some confirmation, it

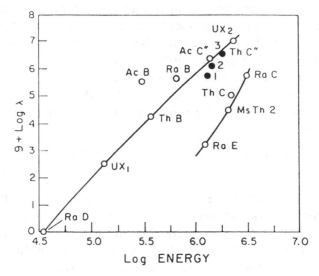

Figure 1.1. Logarithm of the decay constant (inversely proportional to the lifetime) plotted against the logarithm of the maximum decay energy. From Sargent (1933).

was quickly pointed out by Konopinski and Uhlenbeck (1935) that more detailed examination of the spectra showed that his theory predicted too few low-energy electrons and an average electron energy that was too high. They proposed their own model which modified Fermi's theory, but only to the extent that it included the derivative of the neutrino wavefunction rather than the wavefunction itself. They obtained the energy spectrum

$$P(W)dW = G^2 |M|^2 f(Z, W) (W_o - W)^4 (W^2 - 1)^{\frac{1}{2}} WdW$$
(1.3)

which differs from the Fermi prediction by an extra factor of $(W_o - W)^2$ (see Equation 1.2). This predicted more electrons at lower energy and a lower average energy than did Fermi's theory. They cited as support for their modification the spectra obtained from P^{30}, a positron emitter, by Ellis and Henderson (1934) and from RaE (Bi^{210} in modern notation) by Sargent (1932). The RaE spectrum is shown in Figure 1.2 and indicates the superior agreement of their modification with the experimental data. Their model also predicted that $F\tau_o$ would be approximately constant.

They remarked, however, that their improvement did not solve one of the outstanding problems of Fermi's theory. This was that using his theory to explain neutron–proton interactions resulted

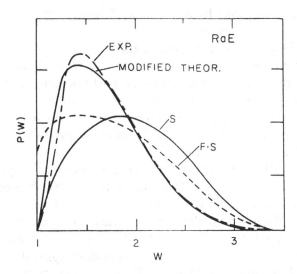

Figure 1.2. Energy spectrum of RaE. *S* is the statistical factor, *FS* is the Fermi formula, and Modified theory is the K-U prediction. Exp is the experimental result of Sargent (1933). From Konopinski and Uhlenbeck (1935).

in predictions that were many orders of magnitude too small.[4] "Our modification of the form of the interaction energy does not of course affect Fermi's explanation of Sargent's law, nor does it change the value of *G* [the coupling constant]. The essential difficulty of the Fermi theory as pointed out by Nordsieck (1934), Tamm (1934), and Iwanenko (1934) therefore remains. Because of the smallness of *G* the interaction between the neutron and the proton through the electron–neutrino field is much too weak to account for the experimental results on neutron–proton scattering" (Konopinski and Uhlenbeck 1935, p. 12). Later work argued that the interactions responsible for nuclear forces and for β decay were distinct.

Konopinski and Uhlenbeck also noted that the theory proposed by Beck and Sitte faced serious difficulties. This theory, which required an electron–positron pair, predicted a symmetric energy spectrum, with only the Coulomb factor $f(Z, W)$ to explain any asymmetry. It disagreed with the experimentally observed simi-

---

[4] Contemporary physicists would not think that a theory of beta decay, a weak interaction, would necessarily apply to the interaction of neutrons and protons, which is a strong interaction, although contemporary theory has unified the weak and electromagnetic interactions and further attempts to include the strong interactions are proceeding.

larity of spectra for both high and low Z nuclei and for both electron and positron emitters.

The Konopinski-Uhlenbeck modification seems to have been accepted by the physics community. In their review article on all of nuclear physics, which remained a standard reference and was used as a student text into the 1950s, Bethe and Bacher, after surveying the evidence, remarked, "We shall therefore accept the Konopinski-Uhlenbeck theory as the basis of future discussions" (Bethe and Bacher 1936, p. 192).[5]

A further modification of Fermi's original theory was proposed by Gamow and Teller (1936). They noted that Fermi had originally required a selection rule of $\Delta J = 0$, and did not include any possible effects of nuclear spin. They included nuclear spin and obtained selection rules of $\Delta J = \pm 1, 0$ for allowed transitions, with no $0 \rightarrow 0$ transitions. This also applied to the Konopinski-Uhlenbeck modification. This required an axial vector form of the interaction, as opposed to Fermi's original vector form. It was realized somewhat later that the tensor form would also work. Gamow and Teller supported their view with a detailed analysis of the decay scheme ThB $\rightarrow$ ThD. "We can now show that the new selection rules help us to remove the difficulties which appeared in the discussion of nuclear spins of radioactive elements by using the original selection rule of Fermi" (Gamow and Teller 1936, p. 897).

The Konopinski-Uhlenbeck theory received further support from the work of Kurie, Richardson, and Paxton (1936), who found that the spectra of N^{13}, F^{17}, Na^{24}, Si^{31}, and P^{32} all fit that model better than did the original Fermi theory. It was in that paper that what was later to be called the Kurie plot made its first appearance. If we look at Equation 1.2 we see that $[P(W)/(f(Z,W)(W^2 - 1)^{1/2}W)]^{1/2}$ is a linear function of W, the electron energy, for the Fermi theory. [In the original papers $P(W)$ is called N, the number of electrons.] For the Konopinski-Uhlenbeck theory the exponent will be $1/4$ for the straight line (see Equation 1.3). Kurie and his collaborators plotted both curves

---

[5] A similar view was expressed by Rose, who remarked, "With these new measurements as a criterion in the case of heavy nuclei and with the results for light nuclei, . . . it seems that the Konopinski-Uhlenbeck theory is capable of accounting for all the features of the continuous β-ray spectrum" (Rose 1936, p. 729).

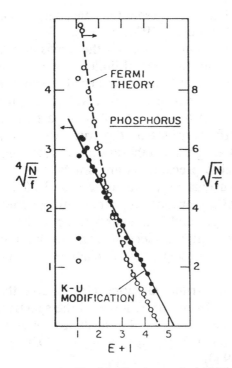

Figure 1.3. "The (black) points marked 'K-U' modification should fall as they do on a straight line. If the Fermi theory is being followed the (white) points should follow a straight line as they clearly do not" (Kurie et al. 1936).

and, as shown in Figure 1.3, the Konopinski-Uhlenbeck theory gave the better fit to a straight line, indicating its superiority.

Kurie's paper also discussed one of the difficulties for the K-U theory. The maximum electron energy, W_o, extrapolated from the straight line graph predicted by the K-U theory seemed to be higher than the experimentally measured value, the value visually extrapolated from the observed energy spectrum. Konopinski and Uhlenbeck had noted this themselves in their original paper. For P^{30} the experimental value of W_o was 6.54 $m_e c^2$, whereas their model predicted 8.0. For Al^{26} the experimental and theoretical values were 6.97 MeV and 8.8 MeV, respectively. It was only for RaE that the experimental and theoretical values agreed at 3.4 MeV. Kurie et al. found that for N^{13} the K-U theory gave a value for W_o of 1.45 MeV whereas the value calculated from known reaction energies was 1.5 MeV. "The excellent agreement of these two values of the upper limits is regarded as suggesting that the high K-U limits represent the true energy changes in a β disintegration" (Kurie et al. 1936, p. 368).

The allowed energy spectrum approaches the x-axis at a small angle making it plausible that the actual limit could differ from the visual one. As Langer and Whitaker remarked: "Experimental difficulties arose because the distribution curve approaches the energy axis gradually and, if the number of beta particles near the end point is not sufficiently great, the true effect may be masked by the natural background inherent in all detecting devices. Moreover, if suitable precautions are not taken, the distribution will have a spurious tail, due to scattered electrons, which approaches the axis asymptotically to much higher energies than the end point" (Langer and Whitaker 1937, p. 713). Scattering of the electrons can result in electron energies that appear to be too high if magnetic focusing or cloud chamber techniques are used.

Much of the additional experimental support for the K-U theory during the late 1930s came from observations of the spectrum of RaE, in which the K-U theory appeared to fit the observed spectrum better than that of Fermi (Watase and Itoh 1938; Flammersfeld 1939; Neary, 1940).

This experimental support for the K-U theory was not, however, completely unambiguous. Richardson (1934) had pointed out earlier the experimental difficulties attendant on measuring the low-energy end of the electron spectrum, including the problem of energy losses due to scattering in the source, which tended to produce too many low-energy electrons. Sargent (1932), too, had commented on the different results obtained from using each of the three main methods of measuring β-decay spectra: range measurements, magnetic focusing, and cloud chamber photographs. Even in the method he chose, that of range in aluminum, he found that the two standard references (Varder 1915; Madgwick 1927) disagreed considerably. O'Conor (1937) pointed out that "Since the original work of Schmidt in 1907 more than a score of workers have made measurements on the beta-ray spectrum of radium E with none too concordant results." In particular, the high-energy end points obtained differed by more than a factor of 2.5. By 1940, however, a consensus seems to have been achieved. As Townsend put it, "the features of the β-ray spectrum of RaE are now known with reasonable precision" (Townsend

1941, p. 365). One feature of these experiments of the late 1930s was the almost universal use of the magnetic spectrometers to measure β-decay spectra. This technique provided both greater precision and accuracy in measuring the electron energy as well as better statistical precision in the spectrum.

As discussed earlier, a major difficulty for the K-U theory was the discrepancy between the measured maximum electron energy and that extrapolated from the theory. This discrepancy became more severe as measurements became more precise. As early as 1937, in their comprehensive review of experimental nuclear physics, Livingston and Bethe remarked, "Kurie, Richardson, and Paxton, have indicated how the K-U theory can be used to obtain a value for the theoretical energy maximum from experimental data, and such a value has been obtained from many of the observed distributions. On the other hand, in those few cases in which it is possible to predict the energy of the beta decay from data on heavy particle reactions, the visually extrapolated limit has been found to fit the data better than the K-U value. The fact that the shape of the distribution curves is checked over wide ranges suggests, however, that the K-U theory has some significance, and it is recorded in the tabulations to follow wherever observed" (Livingston and Bethe 1937, p. 357). Because of this uncertainty they used the visual value to determine atomic masses. Lawson illustrated the difficulty of obtaining unambiguous results in his discussion of the history of spectrum measurements of P^{32}.

The energy spectrum of these electrons was first obtained by J. Ambrosen (1934). Using a Wilson cloud chamber, he obtained a distribution of electrons with an observed upper energy limit of about 2 Mev. Alichanow et al. (1936), using tablets of activated ammonium phosphomolybdate in a magnetic spectrometer of low resolving power, find the upper limit to be 1.95 Mev. Kurie, Richardson, and Paxton (1936) have observed this upper limit to be approximately 1.8 Mev. This work was done in a six-inch cloud chamber, and the results were obtained from a distribution involving about 1500 tracks. Paxton has investigated only the upper regions of the spectrum with the same cloud chamber, and reports that all observed tracks above 1.64 Mev can be accounted for by errors in the method. E. M. Lyman (1937) was the first investigator to determine accurately the spectrum of phosphorus by means of a magnetic spectrometer. The upper limit of the spectrum which he has obtained is 1.7 ± 0.04 Mev (Lawson 1939, p. 131).

Lawson found a value of 1.72 MeV, in good agreement with that found by Lyman. Lyman had already noted that for both P^{32} and RaE the end point obtained by extrapolating the K-U plot was 17 percent higher than that observed.

Similarly, the spectrum of N^{13}, which Kurie had cited in support of the K-U theory, was found by Kikuchi et al. (1939) and by Watase (1940) to be consistent with the Fermi theory if the decay was complex (the parent nucleus, N^{13}, decayed to several different states of the daughter nucleus, C^{13}). Lyman (1939) demonstrated that the spectrum was indeed complex. Although Lyman's results were published in 1939, Watase, in 1940, was not aware of them.

Another developing problem for the K-U theory was that its better fit to the accepted RaE spectrum required a finite mass for the neutrino. This was closely related to the problem of the end-point energy discussed earlier. In fact, the mass of the neutrino was calculated from the difference between the observed end-point energy and the extrapolated end-point energy from the K-U theory. Watase and Itoh (1938) required an m_ν of 0.4 m_e and cited other values of 0.3 by Alichanian, 0.45 by Lyman, 0.5 by Flammersfeld, and 0.52 by Langer. Other work using mea-surements of nuclear masses and reaction energies set limits for the neutrino mass of (0.11 ± 0.31) m_e, (0.002 ± 0.10) m_e, and less than 0.10 m_e (Cockcroft and Lewis 1936; Bonner et al. 1936; Haxby et al. 1940). These results cast serious doubt on a theory that seemed to require a finite neutrino mass. In addition, Ali-chanian, Alichanow, and Dzelepow (1938) and Alichanian and Nikitin (1938) found that the K-U theory fit the spectra of RaE and ThC better but required a neutrino mass of $0.3m_e$ for RaE and $0.8m_e$ for ThC. The difference in mass made it highly unlikely that it could be attributed to a single particle.

Toward the end of the decade, other experimental evidence began to favor Fermi's theory over the K-U modification. Paxton (1937) had noted that the spectra of N^{13} and F^{17} indicated that the experimental upper limit for the electron energy was to be preferred to the extrapolated K-U value and that the K-U model seemed to give too high a value for RaE. Tyler found that for Cu^{64}, for both positrons and electrons (the nucleus decayed into either) the Fermi theory fit the thin source data better than the K-U theory. "The thin source results in much better agreement with the original Fermi theory of beta decay than with the later

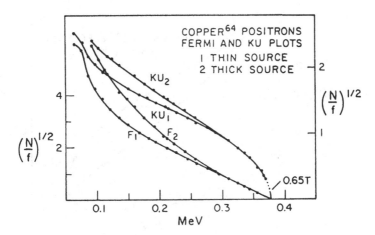

Figure 1.4. Fermi and K-U plots of positrons from thick and thin Cu^{64} sources. From Tyler (1939).

modification introduced by Konopinski and Uhlenbeck. As the source is made thicker there is a gradual change in the shape of the spectra which eventually brings about better agreement with the K-U theory than with the Fermi theory" (Tyler 1939, p. 125). This is shown in Figure 1.4. Similar results were obtained for phosphorus, sodium, and cobalt by Lawson.

In the cases of phosphorus and sodium, where the most accurate work was possible, the shapes of the spectra differ from the results previously reported by other investigators in that there are fewer low energy particles. The reduction in the number of particles has been traced to the relative absence of scattering in the radioactive source and its mounting. The general shape of the spectra is found to agree more satisfactorily with that predicted from the original theory of Fermi than that given by the modification of this theory proposed by Konopinski and Uhlenbeck (Lawson 1939, p. 131). (See Figure 1.5.)

Perhaps the most convincing evidence was provided by Lawson and Cork in their study of In^{114}. Their Kurie plot for the Fermi theory is shown in Figure 1.6 and is clearly a straight line. They pointed out, "However, in all of the cases so far accurately presented, experimental spectra for 'forbidden' transitions have been compared to theories for 'allowed' transitions. The theory for forbidden transitions has not been published" (Lawson and Cork 1940, p. 994). In^{114} was an allowed transition, which allowed a valid comparison between theory and experiment to be made.

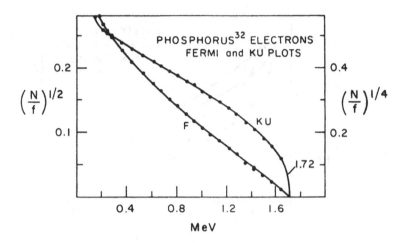

Figure 1.5. Fermi and K-U plots for electrons from phosphorus. From Lawson (1939). The better theory is that which gives the better fit to a straight line.

Similar cautions concerning this type of comparison had been made by Langer and Whitaker (1937) and by Paxton (1937), but not much attention seems to have been paid to them.

It seems clear, in retrospect, that the experimental support for the K-U modification had several problems. The first was the experimental artifact of the excess of low-energy electrons caused by scattering and energy loss in a thick source and the general problem of accurately measuring the high-energy end of the spectrum. The second was using a theory that did not apply to the experimental results.

The calculation of the spectrum shape for forbidden transitions for the Fermi theory was undertaken by Konopinski and Uhlenbeck (1941). They recognized that the new experimental results on both spectra and on the maximum electron energy had removed the basis for their criticism of Fermi's original theory. They noted that there was no a priori reason to expect that the data for forbidden transitions would fit the theoretical prediction for allowed transitions. They calculated the spectral shape for various types of forbidden transitions (i.e., first and second forbidden). In contrast to the spectra of allowed transitions, which were independent of the type of interaction (S, V, T, P, A), they found that the forbidden spectra did depend on the type of interaction. Thus, the shape of forbidden spectra could be used to decide which

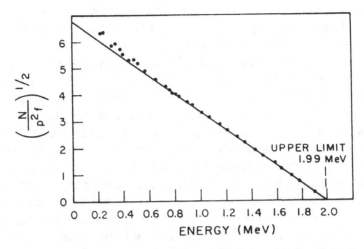

Figure 1.6. Kurie plot for electrons from In[114]. The straight line is the Fermi prediction. From Lawson and Cork (1940).

interaction was responsible for the decay, assuming it was a single type. They applied their new calculation to the spectra of P^{32} and RaE, both thought to be second forbidden decays, and found that they could be fit by either a vector or a tensor interaction. They preferred the tensor theory because that led to the Gamow-Teller selection rules, which seemed to be favored by experimental data. They also noted,

The one encouraging feature of the application of the theory to the experiments is that the decided deviation of the RaE from the allowed form can be at all explained by the theory. . . . *The theory gives a correction factor approximately proportional to* $(W_o - W)^2$ *for an element like RaE. This accounts for the surprising agreements found by the experimenters between their data and the so-called K-U distribution* (Konopinski and Uhlenbeck 1941, p. 320, emphasis added).

Nature seems to have been mischievous in the case of RaE. This mischief was not over. The spectrum of RaE remained a problem into the 1950s. It is currently believed to be a first forbidden transition, but with a cancellation that makes the spectrum appear to be second forbidden.

Marshak (1942) also noted that forbidden spectra could distinguish between different forms of the interaction. Moller (1937) suggested that one could obtain further information from a study of K-electron capture in nuclei. This is a process in which an orbital electron is captured by a nucleus, changing a proton into a neutron, with the emission of a neutrino.

There was further experimental evidence in support of the general features of Fermi's theory in experiments that gave evidence in favor of the existence of the neutrino. In 1936, Leipunski studied the recoiling nuclei during β decay and noted, "The only conclusion that may be drawn is that these results are in favor of the emission of neutrinos during β-decay" (Leipunski 1936). Crane and Halpern (1939) observed both the recoiling nucleus and the decay electron in β decay. They found that momentum was not conserved, arguing in favor of the existence of the neutrino, or, at least, in favor of another neutral particle being emitted during the decay. They were also able to calculate the angle between the decay electron and the neutrino. The Fermi theory and the K-U modification made different predictions for this distribution. The data, although sparse, tended to favor Fermi, but the evidence was far from conclusive. This was the earliest example of an electron–neutrino angular correlation experiment, which would be so important later in efforts to determine the form of the weak interaction. Allen also examined recoil nuclei and concluded, "the recoils were caused by the emission of a neutrino and not by the emission of a gamma ray" (Allen 1942, p. 692).

There were other theoretical developments during this period. In 1935, Yukawa (1935) had suggested that a new particle, of mass intermediate between that of the electron and that of the proton, was responsible for the nuclear force. Observations by Anderson and Neddermeyer (1936) and by Street and Stevenson (1937) had confirmed the existence of such a particle,[6] although it was not until after World War II that it was shown that this was the muon, the decay product of Yukawa's pion. Yukawa (Yukawa, Sakata, and Taketani 1938) speculated that this meson would decay. Although evidence for such decay did not appear until 1940 (Williams and Roberts), attempts were made to incorporate this meson into the theory of β decay by a two-step process in which a nucleon emitted a meson, followed by a meson decay into an electron and a neutrino. It was shown by Bethe

---

[6] See Galison (1987, ch. 3) for details.

(1940a, 1940b), however, that one could not quantitatively fit all the known facts about mesons, nuclear forces, and β decay into a single theory. Konopinski (1943) also showed that the simplest generalization of Yukawa's meson field theory of β decay yielded interactions which were arbitrary linear combinations of S and V, of V and T, of T and A, and of A and P. Symmetry considerations also entered the discussion. In the original Fermi theory the electron and the neutrino were treated differently from the neutron and proton. Critchfield and Wigner (1941) and Critchfield (1943) suggested that they be treated in the same way. They found that although they could not form a completely symmetric interaction, the antisymmetric sum of the scalar, axial vector, and pseudo-scalar interactions was satisfactory.

In 1943 Konopinski published a detailed review of β decay. In that review he summarized the arguments in favor of the original Fermi theory over the Konopinski-Uhlenbeck modification, including the spectra and the maximum energy of decay. *"Thus, the evidence of the spectra, which has previously comprised the sole support for the K-U theory, now definitely fails to support it"* (Konopinski 1943, p. 218, emphasis added). By this time it had also been realized that there were small differences between the two competing theories concerning the $F\tau_0$ values, although they had been originally thought to yield the same conclusion. Fermi theory yielded $F\tau_0$ as a constant, whereas the K-U modification suggested that $(W^2_0 - 1)F\tau_0$ would be more constant. By 1943 the evidence favored Fermi. Similarly the evidence on K-capture, although not conclusive, went against K-U. In the K-capture processes Be^7 to Li^7, K-U theory predicted a ratio of approximately twenty-three for the decay to the ground state as compared to the decay to the excited state. The experimental result was approximately nine. The failure to observe the low-energy peak in the β decay of light elements, predicted by K-U, also argued against it. Konopinski also pointed out that the evidence seemed to support Gamow-Teller selection rules over the original Fermi rules and, in particular, the tensor interaction. Surprisingly, Konopinski made no mention of Fierz's work. Fierz (1937) had shown that if both the S and V interactions were present in the β-decay interaction, or both T and A, then there would be an interference term in the allowed β spectrum, which vanished in the absence of such admixtures. The failure to observe such Fierz interference was cited by later writers as a major step toward deciding the form of the β-decay interaction. It does not appear to have been cited at all in this period.

Thus, at the end of World War II, the situation could be summarized as follows. There was strong support for the Fermi theory of β decay, with some preference for Gamow-Teller rules and the tensor interaction.

2

Toward a universal Fermi interaction; muons and pions

During the 1940s and 1950s the search for a universal theory of weak interactions involved not only nuclear β decay but also the study of mesons. It was some time before it was realized that there was not one meson but two; the pion, which was the particle Yukawa had hypothesized to explain the nuclear force, and the muon, which was its decay product.[1]

We discussed earlier the failure of the attempts to incorporate the presumed decay of Yukawa's meson into the theory of β decay. Until 1940, there was, in fact, no evidence that the meson decayed at all.[2] Although later scientists would interpret some early cloud chamber photographs as evidence for such a decay, this was possible only after the decay had been established.[3] The first definitive evidence for such decay was presented by Williams and Roberts (1940). They presented one cloud chamber photograph that clearly showed a meson stopping in the gas followed by an emerging electron track (Figure 2.1). There was also indirect evidence for decay in the experiments on the anomalous absorption of mesons by Rossi and Hall (1941) and by Neilsen and

---

[1] In the following discussion I will use the term meson for the intermediate mass particle, rather than the modern terms muon and pion. In discussing the history after the 1947 discovery of the pion and the realization that there were two mesons I shall use the modern names.

[2] Kunze (1933), Neddermeyer and Anderson (1938), Maier-Leibnitz (1939), and Nishina, Takeuchi, and Ichimiya (1939).

[3] See, for example, Kunze (1933) and Neddermeyer and Anderson (1938). Neddermeyer and Anderson found three drops in a straight line at the end of a stopped meson track and remarked that it might be evidence for decay. They admitted, however, that the weak light used made it extremely difficult to photograph electron tracks and that such a conclusion was quite uncertain. In Rossi's 1939 review of meson decay he stated, "There is no evidence that the mesotrons which have been stopped actually disintegrate into an electron and a neutrino" (Rossi 1939, p. 296).

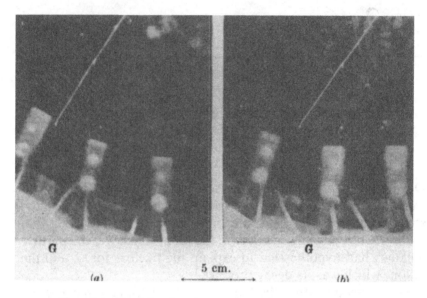

Figure 2.1. Cloud chamber photograph showing muon decay. From Williams and Roberts (1940).

associates (1941). In these 1941 experiments the absorption curves of mesons were measured at two different altitudes. The curves differed from those obtained with an equivalent amount of condensed matter absorber in place of the air. Slower mesons seemed to be preferentially absorbed. This was explained by the decay of the mesons during the extra time needed to traverse the longer distance in air as compared to the time to cross the shorter distance in condensed matter.[4] Neilsen and associates remarked that this was the "first unambiguous proof for the decay effect" (Neilsen et al. 1941). They seemed unaware of the report of Williams and Roberts. The Neilsen group also obtained an estimate for the meson lifetime of $(1.25 \pm 0.3) \times 10^{-6}$ ($\mu c^2/10^8$eV)sec, where μc^2 is the rest energy of the meson. For the then accepted meson mass of approximately $200m_e$, the lifetime is approximately 1.25×10^{-6} sec. In that same year Rasetti (1941) obtained a more direct measurement of the meson lifetime using delayed coincidences

---

[4] The fact that slower mesons decayed preferentially also agreed with the predictions of special relativity. These results also refuted a theory proposed by Fermi that suggested greater absorption in air than in condensed matter, and also predicted that the effect would increase with increasing momentum (Fermi 1940).

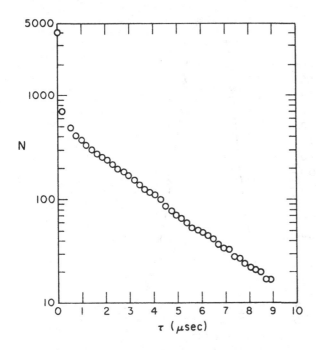

Figure 2.2. Exponential decay curve for the muon obtained by Rossi and Nereson (1942).

between the signal produced by a stopping meson and the signal produced by the decay electron. Using only two points on the decay curve, and an assumed exponential decay, he obtained a value of $(1.5 \pm 0.3)\mu$sec. By 1942, Rossi and Nereson (1942) had shown a definite exponential decay curve (Figure 2.2) and obtained a lifetime of $(2.3 \pm 0.2)\mu$sec.

A rather interesting oddity occurred in these studies of meson decay. De Souza-Santos (1942) reported a 5μsec lifetime that was approximately Gaussian rather than exponential (Figure 2.3). He stated, "We can therefore conclude that the laws responsible for both beta-ray phenomena are altogether different, and that the laws of disintegration of mesons at rest is different from that of beta-rays" (De Souza-Santos 1942, p. 179). As we shall see, this is definitely not the conclusion of our story.

De Souza-Santos's results seem to have been mentioned only once in the literature. Rossi and Nereson (1942) noted that their own work refuted it. Although no explanation of De Souza-Santos's presumed error was offered, the subsequent results show-

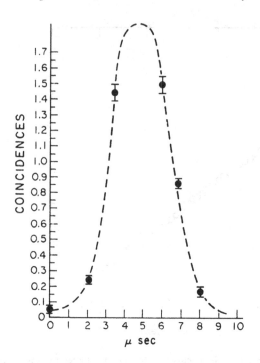

Figure 2.3. Gaussian decay curve for the muon obtained by De Souza-Santos (1942).

ing exponential decay overwhelmed it. A possible explanation appeared in 1949. In a detailed analysis of coincidence circuits Binder (1949) showed that if the resolving time of the circuit was small compared to the Gaussian pulse width, a Gaussian rather than an exponential curve would be observed. De Souza-Santos's paper did not give sufficient detail to decide whether or not this explanation applies and Binder made no mention of that earlier result, but it does seem to be a plausible explanation of this oddity.

By the late 1940s it was generally agreed that the meson decayed with a lifetime of approximately 2.1μsec into an electron and other unknown neutral particles.[5]

There were, however, some puzzling features of meson behavior. As early as 1939, Nordheim and Hebb had remarked, "These

[5] The best measurement using cosmic rays gave a value of (2.15 ± 0.07)μsec (Nereson and Rossi 1943). Using mesons produced by an accelerator Alvarez et al. (1950) obtained a value of (2.09 ± 0.03)μsec.

high cross sections required for the production of energetic mesons seems to imply correspondingly large cross sections for absorption and constitute a serious difficulty for the understanding of their great penetrating power" (Nordheim and Hebb 1939, p. 494). If the mesons were produced copiously then it was hard to understand why so many were observed at sea level. They should have been absorbed by the atmosphere. Tomonaga and Araki (1940) offered an explanation by noting that the Coulomb field of the nucleus would repel the positive mesons and prevent absorption whereas the negative mesons would be attracted to the nucleus and absorbed. "Experimental results are now rather scanty but it does not seem to us merely accidental that all the Wilson tracks which could so far be definitely identified as disintegration electrons are positives, and none of the photographs, in which a negative meson terminates within the cloud chamber, shows a disintegration electron."[6] Further confirmation was provided by experiments that showed an excess of positive mesons (Jones 1939; Hughes 1940). Several groups also measured η, the ratio of the number of decay electrons to the number of mesons stopped in dense material, and found it to be approximately 0.5.[7] This was also in accord with the Tomonaga-Araki theory. If one assumed that the primary radiation consisted of equal numbers of positive and negative mesons then only the positives would decay, giving a value of η approximately equal to 0.5.

The situation changed dramatically when Conversi, Pancini, and Piccioni (1947) reported that whereas negative mesons were absorbed in iron, a substantial fraction of those stopped in carbon decayed. "The results with carbon as absorber turn out to be quite inconsistent with Tomonaga and Araki's prediction" (Conversi et al. 1947, p. 209). Their theory predicted that most of the negative mesons would be absorbed and not decay. This result was analyzed by Fermi, Teller, and Weisskopf (1947). The experimental result implied that the absorption time for negative mesons in carbon had to be of the order of the meson lifetime,

---

[6] Tomonaga and Araki (1940) referred to some of the earlier questionable photographs and also to the definitive picture given by Williams and Roberts (1940).

[7] Rasetti (1941), Rossi and Nereson (1942), and Conversi and Piccioni (1946). Although Auger, Maze, and Chaminade (1941) found $\eta = 1$, the preponderance of evidence favored a value of 0.5.

approximately 10^{-6}sec, to account for the observed decays. Theory predicted, however, that the absorption time should be about 10^{-18}sec, a factor of 10^{12} discrepancy. Something was definitely wrong.

Conversi's results were confirmed by Sigurgeirsson and Yamakawa (1947) for other light materials, and Weisskopf (1947) further underlined the difficulty of reconciling the high rate of meson production with their subsequent weak interaction with matter.

At the Shelter Island conference held during the summer of 1947, Marshak and Bethe offered their two-meson hypothesis as a solution to the difficulty. In their model the Yukawa particle responsible for the nuclear force, which interacted strongly with matter, decayed quickly into another meson that interacted weakly with matter and was the one observed. This explained the copious production but weak absorption. By the time their published version appeared (Marshak and Bethe 1947)[8] the existence of two mesons had been shown experimentally in the cosmic ray cloud chamber photographs of Lattes et al. (1947). Two photographs showed the decay of one particle into another. The masses of the particles, calculated from grain counts, were intermediate between those of the electron and the proton and also differed from each other. "It is therefore possible that our photographs indicate the existence of mesons of different masses" (Lattes et al. 1947, p. 696). Thus were born the pion and muon. This clarified the situation. Work now proceeded separately on the pion and the muon. Both Marshak and Bethe and Lattes and associates suggested that the decay of the pion into the muon and the absorption of muons by nuclei, both weak processes, might originate in the same fundamental interaction. Investigation of the muon concentrated on two problems, the absorption of muons by nuclei and muon decay. It was natural to assume the previous work on η, the ratio between decaying and stopped muons, as showing the effect of nuclear absorption. If nuclear absorption competed with muon decay then $1/\tau^- = 1/\tau^+ + 1/\tau_a$, where τ^-, τ^+, and τ_a are the lifetimes of the positive and negative muons and the ab-

---

[8] A similar model had been proposed by Sakata and Inoue (1946). This work had actually been presented at a symposium in Japan in September 1943, but as they remarked the "printing was delayed owing to the war circumstances."

Table 2.1. *Decay of μ-mesons in several substances*

Substance	Z	τ^- ($\times 10^{-6}$sec)	$f = \tau^- / \tau^+$	f'
H_2O	8	1.89 ± 0.15	0.877 ± 0.069	0.833 ± 0.083
NaF	10.1	1.28 ± 0.12	0.595 ± 0.060	0.602 ± 0.060
Mg	12	0.96 ± 0.06	0.446 ± 0.032	0.558 ± 0.044
Al	13	0.75 ± 0.07	0.34 ± 0.03	0.40 ± 0.03
S	16	0.54 ± 0.12	0.25 ± 0.04	0.28 ± 0.03

Source: From Ticho (1948).

sorption lifetime, respectively. This was not, however, the only possible interpretation.

Epstein, Finkelstein, and Oppenheimer (1948) proposed a theory in which the decrease of τ with increasing Z, the nuclear charge, was due to accelerated decay caused by the increasing electric field of the heavier nuclei. Although the authors themselves regarded the theory as implausible and stated, "Experimental evidence probably disproves theories of this kind," it was consistent with existing data. A test of this hypothesis appeared almost immediately. Wang and Jones (1948) examined muon decays in a cloud chamber that contained both aluminum and lead plates. They rejected the accelerated decay theory because they observed that eight of twenty muons stopped in the aluminum plates showed no decay electrons. The theory required that almost all of them should decay. Marshak later noted:

However, apart from other unrealized predictions, e.g. that photons should accompany μ^+ disintegrations, the "acceleration" theory predicts one decay electron for each μ^- stopped, that is, $f' = 1$[f' is the ratio of decay electrons to stopped muons] although $f < 1$[f is the ratio of the lifetime of the negative muon to that of the positive muon]. If, on the other hand, nuclear absorption of the μ^- meson is the competing process, then the number of decay electrons will be less than the number of μ^- mesons stopping in the material and $f' = f < 1$" (Marshak, 1952, p. 193).

Valley (1947) found $\tau^- = (0.70 \pm 0.06) \times 10^{-6}$sec for aluminum, or $f = 0.32 \pm 0.03$ and $f' = 0.4$. Ticho (1948) obtained values for f and f' given in Table 2.1. Marshak summarized the situation as follows: "It is clear that Valley's and Ticho's results definitely favor the hypothesis of nuclear absorption as the competing process rather than accelerated decay" (Marshak, 1952, p. 193).

These results indicated an absorption lifetime for the muon of approximately 10^{-6}sec in aluminum, a value similar to that of the muon-decay lifetime. This apparent equality of the strength of muon decay, muon absorption, and normal β decay had not escaped the notice of theoretical physicists. They speculated that the three processes were due to the same interaction, the beginning of the search for a Universal Fermi Interaction (UFI). As early as 1947, Pontecorvo remarked:

We notice that the probability (10^6sec^{-1}) of capture of a bound negative meson is of the order of ordinary K-capture [a β-decay process], when allowance is made for the difference in disintegration energy and the difference in the volumes of the K-shell and the meson orbit. We assume that this is significant and wish to discuss the possibility of a fundamental analogy between β-processes and the process of emission or absorption of charged mesons (Pontecorvo 1947, p. 246).

Similar points were made by Puppi (1948) and by Klein (1948). In 1949 Tiomno and Wheeler (1949a) published a detailed theoretical treatment of meson decay. They assumed that the decay process was $\mu^+ \rightarrow \mu_o + e^+ + \nu$, where the μ_o was a neutral particle which could have been another neutrino. They assumed that the decay interaction was of the same form as that of β decay, with the μ^+, μ_o playing the same role as the neutron and proton. They also assumed a single form for the interaction, that is, S, V, T, A, or P. They calculated the decay spectrum for two different assumed muon masses, $200m_e$ and $220m_e$, and for five choices of μ_o mass, 0, $20m_e$, $40m_e$, $60m_e$, and $80m_e$. They also calculated the spectrum for the possible decay process with two neutrinos in the final state, $\mu \rightarrow e + 2\nu$. In this decay there was no natural way to order the decay particles, unlike the case of β decay in which the neutron and proton appeared together as did the electron and neutrino. They considered three cases: (1) the final state is antisymmetric with respect to the two neutrinos ("antisymmetric theory with charge exchange"); (2) in the final state the meson goes to a neutrino and gives charge to a neutrino in a negative energy state, transforming that particle into an electron ("single charge exchange theory"); and (3) in the final state the meson goes to an electron, simultaneously raising a neutrino from a

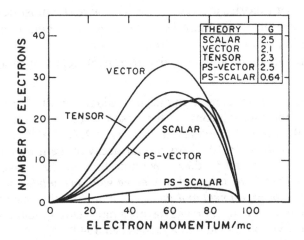

Figure 2.4. "Influence of form of coupling on shape of spectrum for fixed values of the mass of the μ-and μ₀ meson. Contrast this result with the case of ordinary beta-decay, where the atomic nucleus has negligible velocity and the decay curves have the same shape in all five cases" (Tiomno and Wheeler 1949a, p. 148).

negative energy state to a positive energy state ("charge retention theory"). These corresponded to different orders for the meson, electron, and neutrino wavefunctions in the decay Hamiltonians (see Equation 1.1). Because the muon mass is quite different from that of the electron, unlike the approximate equality of the neutron mass and proton mass in β decay, the electron spectrum was not independent of the choice of interaction (see Figure 2.4). Their conclusion was as follows:

It is a remarkable feature of the spectra considered . . . except for the pseudoscalar coupling, that they give coupling constants of the same order of magnitude as those of the corresponding beta-theory. . . . This agreement in order of magnitude between the two *gs* [the coupling constants] is certainly a matter of more than chance and possibly indicates that we have to deal, not with two different theories, but with one and the same theory (Tiomno and Wheeler 1949a, p. 151).

In another calculation they assumed identical coupling (for muon and β decay) and the best available neutron half-life, 30 ± 15 min, and found that the only coupling consistent with both was a vector coupling, although the tensor coupling was possible for the antisymmetrical neutrino theory with charge exchange for muon decay. "In this connection it should be recalled that the vector and tensor theories are the ones which give the most nearly satisfactory account of the main features of nuclear beta decay" (Tiomno and Wheeler 1949a, p. 151).

A second paper dealt with the absorption of negative muons. They stated:

We conclude that it is reasonable to assign a value near 10^{-49} erg cm^3 to this quantity [the coupling constant for meson absorption]. We compared this result with the coupling constants $g_\beta = 2.2 \times 10^{-49}$ erg cm^3 for beta-decay and $g_\mu \approx 3 \times 10^{-49}$ erg cm^3 for decay of the μ-meson on the hypothesis of three end products. We note that the *three coupling constants determined quite independently agree with one another within the limits of error of experiment and theory.* We apparently have to do in all three processes with phenomena having a much closer relationship than we can now visualize (Tiomno and Wheeler 1949b, p. 156–7, emphasis in original).

Lee, Rosenbluth, and Yang (1949) reached a similar conclusion at about the same time.

During this period there was also considerable experimental work on the muon-decay spectrum. In their original muon decay photograph Williams and Roberts (1940) had identified the decay particle as an electron by the ionization produced. Work on the absorption of muons, discussed earlier, assumed that the decay particles were electrons. But, as Hincks and Pontecorvo (1950) were to remark later, "The photograph by Williams and Roberts (as well as others which have been obtained subsequently) showed that a single, lightly ionizing charged particle is emitted in the decay process." It was natural to assume that the particle was an electron, but the conclusion was by no means certain. The question of whether the decay was into two particles, that is, $\mu^+ \rightarrow e^+ + \nu$ or $\mu^+ \rightarrow e^+ + \gamma$, or into three particles, that is, $\mu^+ \rightarrow \mu_o + e^+ + \nu$ or $\mu^+ \rightarrow e^+ + 2\nu$, also remained to be answered. A two-body decay of a muon at rest required a unique energy and momentum for the decay electron. This energy would be approximately one half the muon rest mass, or 50 MeV. During the next few years, experimental work established that the electron emitted in muon decay had a continuous energy spectrum, and that therefore the muon decayed into three particles. It was also shown that the charged decay particle was, in fact, an electron.

Anderson and associates (1947) and Adams et al. (1948) reported two events in which the electron energies were 24 MeV and 25 MeV, respectively. Fowler, Cool, and Street (1948) reported one muon decay with an electron, identified by ionization, of energy 15 ± 3 MeV. Three decay electrons with energies of 13, 18, and 50 MeV were found by Zar, Hershkowitz, and Berezin (1948). They concluded that, "The results here cited would appear to be difficult to reconcile with a monochromatic energy for μ-meson electron decay."[9] The identification by ionization also established that the charge on the decay particle was the same as that of the electron. Thompson (1948), on the other hand, reported nine events with the "preponderance of momenta in the range 40–50 MeV/c," which was what one expected for a two-body decay. Fletcher and Forster (1949) also reported a muon decay with an energy of 28.1 ± 1.5 MeV, calculated from curvature in a magnetic field, and 36 ± 3 MeV, calculated from energy loss in graphite plates. They regarded their energy measurement as less than the 50 MeV required by two-body decay and noted that their result favored decay into more than two bodies. Their energy loss measurements indicated, with a probability greater than 0.95, that the mass of the decay particle was less than 7 m_e.

In 1948 Hincks and Pontecorvo (1948), Sard and Althaus (1948), and Piccioni (1948) searched unsuccessfully for the decay $\mu \rightarrow e + \gamma$. Steinberger, using the absorption of the decay electrons in polystyrene, reported, "the spectrum is either continuous, from 0 to 55 MeV with an average energy ~ 32 MeV or consists of three or more discrete energies. The experiment, therefore, offers some evidence in favor of the hypothesis that the μ-meson disintegrates into three light particles" (Steinberger 1949a, p. 1136). Leighton and associates reported seventy-five cloud-chamber events with an approximately continuous spectrum observed from 9 MeV to 55 MeV and a mean energy of 34 MeV. They stated that they had assumed that "the observed ionizing decay particle is an electron. This, of course, has not actually been proved" (Leighton, Anderson, and Seriff 1949, p. 1436). Their visual estimates of the ionization indicated that the charge of the particles was less than twice that of the electron. This, combined with their momentum measurements, was "therefore wholly consistent with

---

[9] A similar conclusion was reached by Horowitz, Kofoed-Hansen, and Lindhard (1948).

the assumption that the charged particle is an electron" (p. 1436). The continuous spectrum, as well as its shape, also argued in favor of three-body decay. Assuming that the decay produced an electron and two neutral particles, they could calculate an upper limit to the sum of the masses of the neutral particles from the observed maximum electron energy and the known muon mass. They found an upper limit of 30 m_e. "The fact that the production of energetic photons is not observed in mesotron decay leaves open only the possibility that the neutral particles are neutrinos or other neutral particles of low mass. In terms of generally accepted particles, the simplest decay process is thus one which results in the production of an electron and two neutrinos" (Leighton et al. 1949, pp. 1436–7). Assuming that the particles were neutrinos, they calculated the muon mass and found it to be in agreement with the accepted value.

Hincks and Pontecorvo (1949, 1950) reported more extensive work on the decay products in muon decay. They observed that the charged decay particle emitted a γ ray (bremsstrahlung) and concluded that its mass had to be less than 2 m_e. A heavier particle would have been inconsistent with the amount of bremsstrahlung observed. They concluded that the decay particle was indeed an electron. Their own work plus a review of the other existing evidence clearly favored muon decay into an electron and two neutrinos.

In 1951 Sagane, Gardner, and Hubbard (1951) measured the decay spectrum using a magnetic spectrometer. They applied the theoretical analysis of Tiomno and Wheeler to try to decide what was the form of the decay interaction. They obtained a best fit for a tensor coupling in the antisymmetric theory with charge exchange (Figure 2.5). A linear combination of vector and pseudovector couplings could also produce a similar curve. They also concluded that the intensity of decays at the high-energy end of the spectrum approached zero. This was not the result found by Davies, Lock, and Muirhead (1949), Bramson and Havens (1951), Lagarrigue and Peyrou (1951), and Levi Setti and Tomasini (1951). Figure 2.6 shows the nonzero intensity at the maximum energy found in three of these experiments and that of Leighton et al. (1949). Bramson and Havens (1951) also presented further evidence that the decay particle was an electron (actually a positron). In their study of μ$^+$ decays they found six events in which

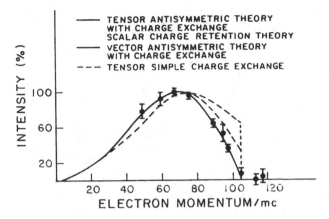

Figure 2.5. Muon-decay spectrum and comparison with theory. From Sagane et al. (1951).

the decay particle disappeared. They attributed this to electron–positron annihilation and concluded, "Hence in addition to having the same charge as the positron, and a mass not much heavier (if at all), this decay particle is now observed to have a third property in common with the positron, that of annihilation" (Bramson and Havens 1951, p. 862).

The most important theoretical analysis of the muon spectrum during this period was that of Michel (1950, 1952). He attempted to analyze the data to find which interaction, or combination of interactions, was responsible for muon decay. He drew no firm conclusion but noted that the interaction suggested earlier by Critchfield and Wigner (1941) to explain β decay, the combination $S - A + P$, was consistent with the data. In addition, this interaction was invariant under any permutation of the order of the four fermions, a pleasing symmetry. There were other attempts by Yang and Tiomno (1950) and by Caianiello (1951) to incorporate symmetry considerations, particle–antiparticle and space reflection symmetry, into the choice of a universal Fermi interaction, but without success. Caianiello rejected the Critchfield-Wigner interaction on the grounds that it disagreed with the zero-intensity end point observed by Sagane. As we have seen, that observation was itself questioned. As discussed later the interaction was also inconsistent with results from nuclear β decay. A note of caution concerning the use of such symmetry principles was sounded by Wick, Wightman, and Wigner (1952). They noted

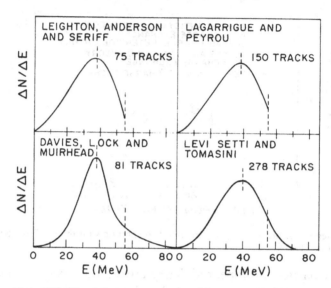

Figure 2.6. Muon-decay spectra obtained by Lagarrigue and Peyrou (1951), Davies, Lock, and Muirhead (1949), Leighton, Anderson, and Seriff (1949), and Levi Setti and Tomasini (1951). From Levi Setti and Tomasini (1951).

that some of these principles had not in fact been tested experimentally, but were assumptions. In later work Michel (1952) showed that an assumed muon decay into an electron and two neutrinos, quite justified as we have seen, leads to a family of spectra described by

$$P(W) \sim (W^2/W_o^4) [3(W_o - W) + 2\rho (4/3W - W_o)], \qquad (2.1)$$

where W is the electron energy, W_o is the maximum electron energy, and ρ is a parameter. As discussed earlier, the theoretical analysis of muon decay is quite complex. The parameter ρ is a function of the particular combination of couplings used (S,A,V,T,P), the way in which the particles are paired in the formulation of the decay interaction, and whether or not the neutrinos are identical. The value of ρ did not uniquely select the correct form of the interaction, although it did serve to reject some combinations, as discussed later.

A clear-cut prediction was, however, obtained from consideration of pion decay. In 1949, Ruderman and Finkelstein (1949) showed "that any theory which couples π-mesons to nucleons also predicts the $\pi \rightarrow (e,\nu)$ decay." They calculated the ratio of the decay of the pion into an electron to its decay into a muon

Table 2.2. *Ratio of $\pi^+ \rightarrow (e, \nu)$ to $\pi \rightarrow (\mu, \nu)$*

Meson	Type of β decay				
	Scalar	Pseudoscalar	Vector	Pseudovector	Tensor
Scalar	5.1	f	f	f	f
Pseudoscalar	f	5.1	f	1.0×10^{-4}	f
Vector	f	f	4.0	f	2.4
Pseudovector	f	f	f	4.0	f

Note: f means forbidden.

for various assumptions concerning the nature of the pion and the nature of the β-decay interaction. Their results are summarized in Table 2.2.

Because pion decay into electrons had not yet been observed they concluded that "the symmetric coupling scheme is in agreement with experimental facts only if the π-meson is pseudoscalar (with either pseudoscalar or pseudovector coupling to the nucleons) and the β-decay coupling contains a pseudovector [axial vector] term" (Ruderman and Finkelstein 1949, p. 1459). A similar result was obtained by Steinberger (1949b).

Although the argument given above requires that the other particle emitted in pion decay be a neutrino, the possibility remained that it could be a photon. That was tested experimentally by O'Ceallaigh (1950). He searched for electron–positron pairs associated with $\pi \rightarrow \mu$ decay which would be produced by a 30-MeV photon (the energy of the photon emitted in such a decay) moving in a direction opposite to that of the muon. He found no such pairs. The probability that such a photon would fail to produce a pair was 4×10^{-3} in his experiment and he concluded that the particle emitted was not a photon. Thus, the evidence favored emission of a neutrino.

Subsequently Friedman and Rainwater (1951) showed that not more than one π^+ meson in 1400 decays into a positron, supporting the conclusion of Ruderman and Finkelstein. F. M. Smith (1951) reported that less than 0.3 ± 0.4 percent of pions did not decay into muons, and Lokanathan, Steinberger, and Wolfe (1954) lowered the limit on the relative probability of electron decay to 2×10^{-5}. This was, in fact, lower than the prediction of

Table 2.3. *Measurements of ρ (1951–6)*

Author	Year	Value of ρ
Lagarrigue	1951	0.43^a
Bramson	1952	0.48^a
Hubbard	1952	0.4^a
Lagarrigue	1952	0.4^a
Vilain	1953	0.50 ± 0.12
Sagane	1954	$0.23^{+0.03b}_{-0.05}$
Vilain	1954	0.50 ± 0.13
Crowe	1955	0.50 ± 0.10
Sargent	1955	0.64 ± 0.10
Sagane	1955	0.22 ± 0.10
Bonetti	1956	0.57 ± 0.14
Sagane	1956	0.62 ± 0.05^c

[a] Corrected for changed muon mass. These were originally reported as 0.2 ± 0.15, 0.41 ± 0.13, 0.26 ± 0.26, and 0.19 ± 0.13, respectively.
[b] A 1951 experiment, using the earlier muon mass, had given $\rho = 0$. In addition, the resoluion was not good enough to show the finite intercept at the maximum energy. The resolution had been improved by 1954.
[c] This result was presented at a meeting of the American Physical Society in 1956. A published paper did not appear until Dudziak, Sagane, and Vedder (1959). In this paper the value of ρ was given as 0.741 ± 0.027. No explanation was offered for the difference between this value and the earlier values.

Ruderman and Finkelstein and remained a problem for several years, and will be discussed in some detail later.

During the first half of the 1950s considerable effort was devoted to measuring ρ. The measurements prior to 1951 suffered from low statistics and/or poor resolution. In addition, as shown later by Vilain and Williams (1954), the determination of ρ was very sensitive to the value of the maximum decay energy or to the mass of the muon. This depended on the method of analysis. In their experiment $\delta\rho = -8\delta W_o/W_o$, where $\delta\rho$ is the change in ρ and δW_o is the change in the maximum energy. During this period the best value for the muon mass changed by 5 percent, which would result in a change in ρ of approximately 40 percent. The results are shown in Table 2.3. As one can see, by 1956 the value of ρ was approximately 0.5 (or perhaps a little higher).

The only measurements that were in disagreement were those of Sagane, Dudziak, and Vedder (1954). They had worried about the problem. "As the discrepancies of the ρ value are quite serious, during the last five months we have made a close examination of the possible systematic errors that we might have overlooked. Our present analysis of our data cannot account for this serious difference in ρ value" (Sagane et al. 1954, p. 864). Nevertheless, the preponderance of evidence was regarded as favoring a larger value of ρ.

The theoretical analysis was provided by Michel and Wightman (1954) although, as Michel (1957) remarked in a review summarizing the situation at the end of 1956, essentially the same analysis applied then. They restricted themselves to considering interactions for muon decay which were combinations of S, T, and P. As will be discussed in detail subsequently, the STP interaction was favored by the evidence from nuclear β decay at this time. The parameters they considered were ρ and λ, essentially the ratio of the $F\tau_0$ values for muon decay and for nuclear β decay. At the time, the best value for λ was 1.16 ± 0.12 and $\rho = 0.5$. They found that only the combination $\pm S + T \mp P$, which predicted $\rho = \frac{3}{8}$ and $\lambda = \frac{4}{3}$, was in even reasonable agreement with the measured values. Other STP combinations were excluded.[10] They also considered a more general interaction of the form $\pm S + T + pP$, where p is a numerical parameter. They found essentially two possible fits. For distinguishable neutrinos emitted in muon decay, $\pm S + T$ gave $\rho = 0.54$, $\lambda = 1.17$, in good agreement with the observations. For indistinguishable neutrinos, $\pm S + T \mp 3.5P$ gave $\rho = 0.58$ and $\lambda = 1.10$.

By 1956 the values of ρ were regarded as being somewhat higher than 0.5[11] and λ was found to be 1.05 ± 0.04 from neutron decay, and 1.11 ± 0.06 from nuclear β decay. These values still excluded those combinations rejected earlier by Michel and Wightman, along with some others that had been proposed in the meantime.[12] It seems fair to say that in 1956 the theory of muon decay was consistent with an interaction composed of S, T, and P.

[10] These combinations had been proposed by: Pursey (1952), Finkelstein and Kaus (1953), Konopinski and Mahmoud (1953), and Peaslee (1953).

[11] Michel and Wightman (1954) cited the values 0.64 ± 0.10 and 0.50 ± 0.10 given by Sargent et al. (1955) and by Crowe, Howe, and Toutfest (1955), respectively.

[12] In addition to the references given in (1957), Michel cited: Pursey (1951), Mahmoud and Konopinski (1952), Pryce (1952), Stech and Jensen (1955), Peaslee (1955), and Tiomno (1955).

3

Beta-decay theory following World War II

As we saw in the last chapter, there was no generally agreed upon theory of muon decay processes in the early 1950s, although the decays were consistent with an *STP* combination. It was known that the coupling constants for those processes were approximately equal to those for nuclear β decay. This suggested the idea of a Universal Fermi Interaction that would apply to all weak interactions. The situation was quite different in the area of β-decay theory, where a consensus existed. In his 1943 review of the theory, Konopinski had noted the general support for Fermi's theory with a preference for the Gamow-Teller interaction involving the tensor (T) or axial vector (A) forms of the interaction. By the time of a 1953 review article by Konopinski and Langer, they stated, "As we shall interpret the evidence here, the correct law *must be* what is known as an *STP* combination" (1953, p. 261, emphasis added). In this section I will examine the evidence and arguments for this definite conclusion, which, as the subsequent history will show, was wrong, but not unreasonable.

There were several attempts, based on general theoretical principles, to reduce the arbitrariness of the forms of the interactions involved in β decay. Recall that Critchfield and Wigner (1941) and Critchfield (1943) had, on the basis of symmetry considerations, found that the antisymmetric sum of the S, A, and P forms would fit the data. Similarly, Tolhoek and de Groot (1951) argued that the emission of electrons and positrons should be indistinguishable, except for electromagnetic effects, in physically equivalent situations (i.e., when the initial and final nuclear states for the two processes are interchanged). They found that unless the Fermi interaction was either a combination of the S, A, and P forms alone, or of the V and T forms alone, there would be differences between the two processes, which they regarded as impermissible. The *SAP* combination was consistent with the Critchfield-Wigner hypothesis. Subsequent experimental work, to be discussed later, refuted both of these hypotheses.

During this period, scientists also began to recognize the importance of Fierz's work (1937) for clarifying the nature of the β-decay interaction. Fierz had shown that if the interaction contained both *S* and *V* or *T* and *A* then certain interference terms would appear in the β-decay spectra. The experimental evidence, particularly the linearity of the Kurie plot discussed earlier (see Figure 1.6), argued against such interference terms. The conclusion that $G_S G_V = G_T G_A = 0$ (the coupling constants) was consistent with both the Tolhoek-de Groot and the Critchfield-Wigner hypotheses.[1]

The presence of the *T* or *A* form of interaction, which gives rise to the Gamow-Teller selection rules, in at least part of the β-decay interaction was shown by Mayer, Moszkowski, and Nordheim (1951). They used decay data from nuclei with odd mass number to support their assignments of nuclear spins and parities. On the basis of these assignments they found approximately twenty-five decays for which $\Delta J = \pm 1$, with no parity change. These could occur only if the *T* or *A* forms were present.

3.1 THE SPECTRUM OF CU^{64}

Fermi's theory also received support from the detailed examination of the shape of allowed β-decay spectra. Of particular interest is the spectrum of Cu^{64}, which emits both electrons and positrons. As discussed earlier, Tyler's (1939) results from Cu^{64} had shown that as the radioactive source was made thinner the observed spectrum approached the prediction of the Fermi theory more closely, whereas the thicker source spectrum fit the Konopinski-Uhlenbeck modification better. This had been one of the major pieces of evidence favoring the original Fermi theory. More precise measurements of the spectrum, however, particularly at low electron energies, revealed problems for the Fermi theory.

Both Backus (1945) and Cook and Langer (1948) found more

---

[1] Note, however, that the interference would not exist if the interaction involved only a single form *S*, *A*, *V*, *T*, or *P*.

electrons and positrons at low energies than the Fermi theory predicted.[2] Cook and Langer found no emitted gamma rays, which eliminated a complex spectrum, one in which the decay goes to two or more final states, as a possible explanation of the discrepancy. If the spectrum were complex such gamma rays should be observed. They also argued that, contrary to Lewis and Bohm (1946), both the positron and electron spectra were in disagreement with the Fermi theory, and rejected their linear combination of interactions as an explanation of Backus's results.

The technical difficulties of performing these experiments were severe. Owen and Primakoff (1948) offered a method of correcting the spectra obtained with a magnetic spectrometer, the most frequently used instrument. They included corrections for distortions due to the focusing properties of the spectrometer, those due to the elastic and inelastic scattering of electrons in the source and its backing, and those due to scattering by the walls, slits, and residual gases. They applied their corrections to the number of positrons and electrons emitted from Cu^{64} and obtained "results in better agreement with the Fermi theory than if there had been no corrections," although a discrepancy still remained.

Longmire and Brown (1949) calculated the effects on β-decay spectra due to screening by atomic electrons, and a smaller correction due to using a relativistic rather than a nonrelativistic approximation to the Coulomb correction factor. They found that the screening correction was too small to account for the deviation from the predicted values seen by Cook and Langer.

The experimental investigation of Cu^{64} was continued by Wu and Albert (1949a), with particular emphasis on the thickness and uniformity of the radioactive source. They found that as the source thickness was reduced from 0.3 mg/cm^2 to ~0.1 mg/cm^2 there

---

[2] Lewis and Bohm (1946) had attempted a theoretical explanation of the discrepancy. They pointed out that although the positron spectrum was in excellent agreement with theory, the electron spectrum was not. This was unlikely, in their view, to be explained on the basis of an instrumental asymmetry. They noted that two particular linear combinations of the forms of the Fermi interaction, the difference between the scalar and vector interactions and the difference between the tensor and axial vector interactions, would improve the fit to the electron spectrum without impairing the positron agreement. They admitted, however, that their model predicted K-capture probabilities much smaller than those observed and too strong an energy variation of decay lifetime. They suggested further work on the low-energy end of the spectra to clarify the situation.

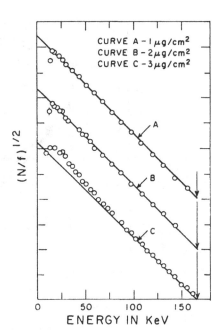

Figure 3.1. The effect of target thickness on the decay spectrum of S^{35}. From Albert and Wu (1948).

was a gradual but consistent reduction in the deviation at low energies. At ~0.1 mg/cm^2 the deviation was smaller than any that had been previously reported. (See Figure 3.1 for the effect of target thickness on the spectrum of S^{35}, discussed in the next section.) They also applied the corrections calculated by Longmire and Brown and found that they obtained good agreement down to an energy of about 70 KeV. They speculated that, in light of the decrease in the deviation due to a thinner and more uniform source and a more rigorous Coulomb correction, the spectrum really did fit Fermi's theory and that the remaining small discrepancy was due to instrumental effects. A second paper (Wu and Albert 1949b) pointed out that Longmire and Brown had overlooked a significant factor in their calculation that affected the positrons more than the electrons. Although it did not account for all the observed deviation in the low-energy region it did improve the overall agreement. They noted, "When the corrections are applied, the agreement between the experimental and theoretical values is excellent," and concluded, "The Fermi theory probably does approximate the true distribution for negatrons

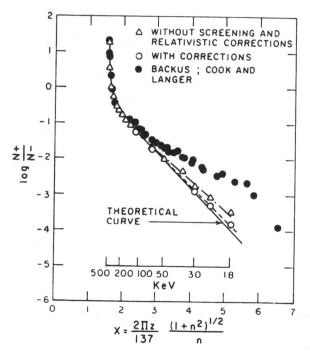

Figure 3.2. The logarithm of N_+/N_-, the ratio of positrons to electrons as a function of energy, for Cu[64]. From Wu and Albert (1949b).

and positrons at low energies." Their results are shown in Figure 3.2 and clearly indicate an improved agreement between theory and experiment when compared with Backus's results, which are also shown.

Owen and Cook (1949a), however, did not believe that the discrepancy at low energies could be explained by energy loss in a thick source or by the failure to make the appropriate corrections to the spectrum. In their work on Cu[61], another isotope of copper, they prepared an extremely thin source that "*could not be seen visually and could only be detected by its activity* [emphasis in the original]." The discrepancy at low energies persisted and they stated that they found it difficult to account for this by instrumental distortions. The effect in Cu[61] was later explained by the observation of the complexity of the decay spectrum, discussed earlier. The gamma rays that accompany such a decay were found almost immediately by Boehm's group (1950) and by Owen and Cook, themselves (Owen, Cook, and Owen 1950).

Langer, Moffat, and Price (1949) found that their sources, although uniform in appearance, had variations in intensity as large as 100 to 1. When they used a new, thin, uniform source they found a deviation much smaller than any that had been previously observed. They expected even better agreement when they applied the correction of Longmire and Brown to their data, which they had not as yet done. They concluded, "It appears, then, that there is no longer any real disagreement between the experimental spectrum shape and that predicted for an allowed transition by the Fermi theory" (Langer et al. 1949, p. 1726). Owen and Cook (1949b), in the adjoining article, reported linear results in their Kurie plot down to less than 50 KeV and suggested that the previous deviations were due to finite source thickness.

We see, once again, the difficulty of comparing theory and experiment. There were numerous technical difficulties in performing these experiments including the thickness and uniformity of the sources and distortions due to the spectrometer. Until these were corrected the experimental results were in error. There were, in addition, corrections that had to be applied to the theoretical spectrum. Only when both the experimental difficulties had been corrected and the theoretical corrections applied could a valid comparison between theory and experiment be made.

3.2 THE SPECTRUM OF S^{35}

Further evidence in support of the Fermi theory was provided by measurements of the spectrum of S^{35}. Here, too, there were experimental difficulties involving source thickness. Early experimental work found deviations from the Fermi theory at low energies (Cook, Langer, and Price 1948a, b). Later experiments found that as the target thickness was reduced these discrepancies disappeared (see Figure 3.1) (Albert and Wu 1948; Langer, Motz, and Price 1950). The issue seems to have been resolved when Gross and Hamilton (1950) used a new type of electrostatic spectrograph, which had very high resolving power at sufficient intensity. This allowed a precise measurement of the spectrum below 30 KeV. They found excellent agreement with the Fermi theory down to 6 KeV, far lower than had been previously obtained, and attributed the small deviation below that energy to backscattering and absorption in the source and backing. Further evidence was provided by spectrum measurements for the neutron, H^3, He^6, and N^{13}.[3] These results did not have significant problems due to scattering and energy loss in the source because the sources used were gaseous.

The linearity of these Kurie plots both supported the Fermi theory and eliminated the Fierz interference terms. As discussed earlier, Fierz had shown that if the decay interaction contained both S and V terms, or both A and T terms, there would be an energy-dependent term in the spectrum, of the form $1 + a/W$, that would destroy that linearity. This was a significant step toward finding the specific form of the decay interaction.

Evidence in favor of the presence of S or V terms was provided by the work of Sherr, Muether, and White (1949) and Sherr and Gerhart (1952) on the favored decays (short half life) of C^{10} and O^{14}. The argument was as follows. O^{14} was found to decay into an excited state of N^{14}, N^{14*}, which had exactly the energy expected if that state differed from the ground states of C^{14} or O^{14} in no other way than by having one proton more or less. The ground states of C^{14} and O^{14} have spin O and it was inferred that N^{14*} must also have spin O. Thus the transition from O^{14} to N^{14*} was permitted only for the S and V forms of the decay interaction and forbidden for A or T.[4] A similar argument held for the decay C^{10} to B^{10**}.

3.3 FORBIDDEN SPECTRA

We discussed earlier the evidence from the lifetimes of transitions presented by Mayer et al. (1951) that supported the presence of the Gamow-Teller terms (T or A) in the decay interaction. These depended on the correct assignments of nuclear spins and parities which, although quite reliable, still had at least a small element of uncertainty. Evidence in favor of the presence of these terms that did not depend on a knowledge of nuclear spins came from the examination of the spectra of unique forbidden transitions, those for which in an n-times forbidden transition the change in spin $\Delta J = n + 1$. These transitions require the presence of A or T. In addition, only a single form of the interaction makes any appreciable contribution to the decay. This allows the prediction

---

[3] Cook, Langer, and Price (1948c), Hornyak and Lauritsen (1950), Perez-Mendez and Brown (1950), and Robson (1951).
[4] For a detailed discussion see Konopinski and Langer (1953).

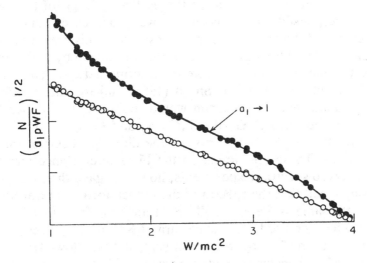

Figure 3.3. The unique, once forbidden spectrum of Y^{91}. The conventional Fermi plot, $a_1 = 1$, does not give a straight line. The correction factor $a_1 = C[(W^2 - m^2_0 c^4) + (W_0 - W)^2]$, calculated by Konopinski and Uhlenbeck (1941) does give a linear plot. Figure from Konopinski and Langer (1953).

of the spectral shape for such a transition. Konopinski and Uhlenbeck (1941) showed that for such an n-times forbidden transition the spectrum would be that of an allowed Fermi transition multiplied by an energy dependent factor $a_n(W)$. For a first forbidden transition $a_1 = C [(W^2 - m^2 c^4) + (W_o - W)^2]$. Langer and Price (1949) observed such a transition for Y^{91} as did Peacock and Mitchell (1949) for Cs^{137}. Figure 3.3 shows the Kurie plot obtained by Langer and Price for a Fermi spectrum and for that spectrum multiplied by a_1. These results demonstrated the presence of T or A terms in the decay interaction.[5]

[5] Mayer, Moszkowski, and Nordheim (1951) listed twenty-five transitions expected to have a spin change of 2 and a parity change, and that would therefore be expected to have this same unique spectrum shape. Konopinski and Langer (1953) reported that seventeen of these spectra had been measured and all had the predicted shape. They listed the following isotopes: Cl^{38}, A^{41}, K^{42}, As^{76}, Kr^{85}, Rb^{86}, Sr^{89}, Sr^{90}, Sr^{91}, Y^{90}, Y^{91}, Y^{92}, Sn^{123m}, Cs^{137}, Pr^{142}, Tm^{170}, and Tl^{204}. Subsequent work also showed the existence of the predicted spectral shapes for second and third forbidden transitions. The spectrum of Be^{10} provided the evidence for second forbidden transitions and that of K^{40} for third forbidden transitions. A comprehensive review was contained in Wu (1950). Other experiments were Fulbright and Milton (1949), Alburger (1950), Alburger, Hughes, and Eggler (1950), Bell and Cassidy (1950), Bell, Weaver, and Cassidy (1950), Feldman and Wu (1950), and Good (1951).

Further progress in isolating the particular forms of the interaction responsible for weak decays was made by examining the spectrum of "once-forbidden" transitions. This spectrum had been calculated for single forms of the interaction by Konopinski and Uhlenbeck (1941). Interference terms between two forms were calculated by A. M. Smith (1951) and by Pursey (1951). They found that the spectrum would contain energy dependent terms of the form $G_V G_T/W$, $G_A G_P/W$, and $G_S G_A/W$ (where the Gs are the coupling constants for the different forms of the interaction). These are similar to the Fierz interference terms in the allowed spectra. Experiments, however, gave the rather surprising result that the spectra of these once-forbidden transitions were the same as that of an allowed transition. In 1946, Siegbahn (1946) had reported that the spectra of Na^{24} and P^{32}, which were first and second forbidden, respectively, had the allowed spectrum shape. More conclusive evidence was provided by Langer, Motz, and Price (1950). Their linear spectrum of Pm^{147}, a once-forbidden transition, demonstrated the absence of the interference terms.

Our earlier discussion of allowed transitions restricted the forms of the interaction to *STP, SAP, VTP,* or *VAP,* or doublets taken from these combinations. The absence of interference terms in the once-forbidden spectra eliminated the *VT, SA,* and *AP* combinations. The *VP* doublet was eliminated because it did not allow the favored Gamow-Teller selection rules and the "unique spectra" discussed earlier. Thus, one was left with either the *STP* triplet or the *VA* doublet. This conclusion also eliminated the Critchfield-Wigner and the Tolhoek-de Groot hypotheses.

3.4 RADIUM E

The spectrum of RaE provided the decisive evidence. The crucial analysis was provided by Petschek and Marshak (1952). Because of its importance in our story, this analysis will be discussed in detail.

We recall that the RaE spectrum was one of the major pieces of evidence favoring the K-U modification over Fermi's original theory. By the early 1940s agreement had been achieved on the spectral shape. The comparison with the calculation by Konopinski and Uhlenbeck of the forbidden spectra had established that the RaE spectrum could be explained by the Fermi theory if the spin change were 2 and the transition did not involve a change in the parity of the nuclear state.[6] Petschek and Marshak argued that the recent successes of the nuclear shell model had made it desirable to reexamine this theoretical fit because the RaE nucleus was ideal for shell model calculations. They found that the transition had to involve a change in parity and a spin change of 0 or 2. This cast doubt on the previous fit. Assuming a change in parity and a spin change of 0 or 2, they attempted to fit the RaE spectrum using all possible linear combinations of the five interaction forms (S, A, V, T, and P). They did not include those excluded by the Fierz condition, SV and TA. Their calculation used the forbidden spectra correction factors of Konopinski and Uhlenbeck (1941) and the A. M. Smith interference formulas (1951). They also made a correction due to the finite size of the nucleus taken from Rose and Holmes (1951).[7] They made no correction for screening by atomic electrons, which was said to be small (Reitz 1950).

Petschek and Marshak found that only a linear combination of the T and P forms of the interaction, with a spin change of 0, could fit the spectrum. They noted, however, that the their theoretical fit was quite sensitive to their assumptions. "Thus, an error in the finite radius corrections of approximately 0.1 percent leads to an error of up to 25% in $C_{1(T+P)}$ [the theoretical correction term]." This was because of the interference and almost complete cancellation of several large terms in the theoretical expression for the spectrum.

They concluded, "Within the errors noted previously, the linear combination of tensor and pseudoscalar interactions corresponding to $\Gamma = 13 \pm 1$ can be regarded as giving a satisfactory fit of the RaE spectrum.[8] Moreover, it is the only linear combination which can explain the forbidden shape of the RaE spectrum if the spectrum is simple and if the parity prediction of the shell model is accepted. Subject to this qualification, our calculation provides the first clear-cut evidence for an admixture of the pseu-

---

[6] This explanation also required that the spectrum be simple, i.e., no gamma rays. There was good experimental evidence for this from the spectral measurements.

[7] Rose and Holmes noted, "The only well-investigated case for which the finite size corrections should be important is the RaE spectrum."

[8] Γ was the ratio between the two different interaction terms in the spectrum, and was in rough agreement with the estimated value of those terms.

doscalar interaction to explain all β-ray phenomena" (p. 698). This was the only evidence that led Konopinski and Langer to choose the *STP* combination rather than the *VA* combination.

Once again, however, the spectrum of RaE proved to be an unfortunate piece of evidence on which to base a theory choice. Part of the problem was due to the theoretical corrections to the experimental spectrum. The finite radius corrections of Rose and Holmes, used by Petschek and Marshak in their analysis of the RaE spectrum, used only the first few terms of a power series expansion. The validity of this procedure depended on the electron de Broglie wavelength being large compared to the nuclear radius. Rose and Perry (1953) and Rose, Perry, and Dismuke (1953) pointed out that when the electron momentum and the charge of the nucleus were large this assumption was no longer justified. They used a computer calculation to give an essentially exact result, which no longer depended on the approximation.

Yamada (1953) used these calculations in a theoretical reinvestigation of the RaE spectrum and concluded that the *ST* or *VA* combinations of interactions, with a spin change of one, fit the observed spectrum as well as the *TP* combination with no spin change.[9] In addition, the analysis of Petschek and Marshak required a very large amount of pseudoscalar interaction. This created problems elsewhere, such as in the high-energy end of the allowed spectra (Ahrens, Feenberg, and Primakoff 1952).

Rose and Osborne (1954) provided a new, detailed analysis of the pseudoscalar interaction and found a correction factor quite different from that used by Petschek and Marshak. When they applied it to the spectrum of RaE they found that they could not fit the data with the assumption that the spin of RaE was zero. They suggested that the spin was 1. They cited as independent evidence Yamada's fit to the spectrum using spin 1, the decay chain RaD → RaE → RaF, and the measurement of the hyperfine spectrum of RaE, all three of which were consistent with spin 1.

All of this theoretical analysis became moot when K. Smith measured the spin of RaE directly, using an atomic beam method, and found a value of one.[10] This removed the only evidence sup-

---

[9] Other combinations also fit the data, but these were regarded as implausible because they also required corrections to the nuclear matrix elements.

[10] I have not been able to find a published reference to this measurement. Wu (1955a) cited it as a private communication in her talk at the 1954 Glasgow

porting the presence of the pseudoscalar (*P*) interaction in the theory of β decay.

The demise of the RaE evidence removed only the necessity of including the pseudoscalar interaction in the theory of β decay. The *STP* combination remained the preference of most of the physics community. This was because the evidence from angular correlation experiments, to be discussed in detail in the next section, agreed with *S* and *T* rather than with *V* and *A*.

3.5 ANGULAR CORRELATION EXPERIMENTS

Angular correlation experiments are those in which both the decay electron and the recoil nucleus are detected in coincidence. The experiments measured the distribution in angle between the electron and the recoil nucleus, for a fixed range of electron energies, or measured the energy spectrum of either the electron or the nucleus, at a fixed angle between them.[11] As we shall see, these quantities are quite sensitive to the form of the decay interaction and became one of the decisive pieces of evidence in the search for the form of the interaction.

These experiments were extremely difficult to perform correctly. In the case of solid sources the problems of energy loss and scattering in source, which could change the correlation, were exacerbated by the very low energy of the recoil nucleus. This tended to favor the use of gaseous sources in which the scattering and energy loss were minimized. Such sources had problems with geometry (the gas would fill the available volume making it difficult to define the decay volume precisely, a point we shall return to later) and with obtaining sufficient intensity. After World War II, the cyclotrons and nuclear reactors needed to produce sufficient quantities of these radioactive gases became available. Even for gaseous sources accurate experiments required the use of a monatomic gas. If the gas molecules were diatomic the charge could appear on the stable atom, giving rise to an inaccurate dis-

---

conference and in her review article (1955b) in Siegbahn (1955). Siegbahn's book was a standard reference work for nuclear physicists. This work was also cited as a private communication in the standard reference on molecular beam physics, Ramsey (1956).

[11] For an interesting discussion of these experiments and their status in the mid–1950s see Kofoed-Hansen (1955).

tribution. Thus, as we shall see, most of the important experiments were done with monatomic noble gases. The detection of the electrons and, in particular, the recoil nucleus was quite difficult. Although there were improvements in such detectors in the postwar period, we find that even into the 1950s individual experimenters were growing their own scintillating crystals, making their own scintillating plastic, and constructing slow-ion detectors.[12] An indication of the difficulty of getting reliable results from such experiments is shown in the work of Sherwin on P^{32}, a solid that he produced in a monolayer. In a series of experiments he reported that the best fit to the distribution in Θ, the angle between the electron and the neutrino, was $(1 - \beta \cos\Theta)$ (Sherwin 1948a), where $\beta = v_e/c$, $(1 + \beta \cos\Theta)$ (Sherwin 1949), and $(1 + \cos\Theta)$ (Sherwin 1951), respectively. This angle could not be measured directly, but it could be calculated from measurements on the electron and the recoil nucleus. In an experiment on Y^{90}, Sherwin reported that $(1 - \beta \cos\Theta)^2$ was the best fit to the angular correlation [Sherwin, 1948b].

The earliest experiments were designed primarily to demonstrate the existence of the neutrino by showing that momentum was not conserved in β decay. If the electron and the recoil nucleus did not have equal and opposite momenta then there must be another neutral particle emitted in the decay to take away the remaining momentum.[13] We discussed earlier the experiments of Leipunski (1936), Allen (1942), and Crane and Halpern (1939). We also saw that the latter experiment, albeit with extremely limited statistics, favored the Fermi theory over the K-U modification.

In the late 1940s these experiments began to assume major importance in determining the form of the β-decay interaction. Hamilton (1947) calculated the form of the angular distribution to be expected for both allowed and forbidden decays, if the decay involved only one type of interaction (S, V, T, A, P). He found, for allowed transitions, that the angular distributions would be:

---

[12] Private communication from Brian Ridley, one of the physicists involved in these experiments.

[13] A survey of some of these early experiments is contained in Crane (1948). Although earlier in this century the violation of conservation of energy and momentum in β decay was considered as a possibility, by this time the experiments were regarded as showing the existence of the neutrino.

Scalar $(1 - \beta \cos \theta)$
Vector $(1 + \beta \cos \theta)$
Tensor $(1 + \frac{1}{3} \beta \cos \theta)$
Axial vector $(1 - \frac{1}{3} \beta \cos \theta)$
Pseudoscalar $(1 - \beta \cos \theta)$

A more general treatment of this problem was provided by de Groot and Tolhoek (1950). They found that the general form of the angular distribution for allowed decays was $(1 + \alpha\beta \cos\Theta)$, where α depended on the combinations of the particular forms of the interactions contained in the decay Hamiltonian. Their results for single forms agreed with those of Hamilton.

One of the most important angular correlation experiments was on the decay of He^6. This decay was a pure Gamow-Teller transition and thus was sensitive to the amounts of A and T present. This experiment was first performed by Allen, Paneth, and Morrish (1949). Their experiment measured the energy spectrum of the recoil nucleus when the angle between the electron and the recoil nucleus was 180° and when it was 162°. They found, for both angles, that the best fit to the angular correlation was $(1 - \frac{1}{3} \beta \cos\Theta)$. This result was, however, uncertain. They concluded, "The results of this experiment seem to rule out the 'no neutrino' assumption and give some indication of agreement with the $(1 - v/3c \cos\Theta)$ correlation predicted by the axial vector form of interaction which follows Gamow-Teller selection rules" (Allen, Paneth, and Morrish 1949, p. 577).

This was not, however, the result reported by Rustad and Ruby (1953). This 1953 experiment was regarded by the physics community in the mid-1950s as establishing that the Gamow-Teller part of the decay interaction was predominantly tensor (T). Several review papers on the nature of β decay in this period stated this conclusion rather emphatically.[14] In addition, as we shall see later, other experiments assumed that the Gamow-Teller interaction was tensor, based on this result.

Because of its importance in this history, I will discuss this experiment in detail. A schematic view of the experiment is shown

[14] Wu (1955a), Kofoed-Hansen (1955), and Ridley (1956). Ridley (p. 215) pointed out that the interpretation of the Allen, Paneth, and Morrish (1949) result was made difficult because of technical problems with the apparatus (p. 215). He also regarded the later results of Rustad and Ruby and of Allen and Jentschke as more reliable.

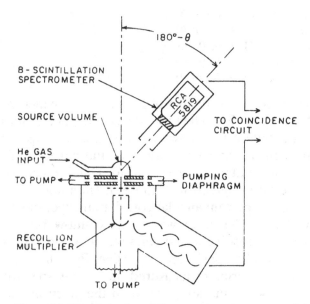

Figure 3.4. Schematic view of the experimental apparatus of the He⁶ angular correlation experiment of Rustad and Ruby (1953, 1955).

in Figure 3.4. The decay volume, extremely important for measuring the angular correlation, was defined by a 180-microgram/cm² aluminum foil hemisphere and the pumping diaphragm. The direction and energy of the electron were determined by a stilbene scintillation spectrometer, which could be positioned at any angle between 100° and 180° from the direction of the recoil nucleus. The recoil nuclei were detected by an electron multiplier and a coincidence circuit selected events in which the recoil detector was activated from 0.08 to 1.15 microseconds after the electron scintillation counter. This was to take into account the longer time of flight of the heavier recoil nucleus, which varied with the energy of the nucleus. Scattering in the helium gas, which might have distorted the angular correlation, was reduced by having the apparatus continually evacuated, and steps were taken to prevent the pump oil from returning to the apparatus.

Two different results were reported. The first was the coincidence rate as a function of the angle between the electron and the recoil nucleus, for electrons with total energy in the range 2.5–4.0 mc^2. The second was the energy spectrum of the electrons when the angle between the electron and the recoil nucleus was

180°. Both results are shown in Figure 3.5. The calculated values of these curves for the axial vector (A) and tensor (T) forms of the decay interaction are also shown. The dominance of the tensor form is clear. Rustad and Ruby concluded, "The consistency of both experiments makes it unlikely that instrumental discrimination resulting from the variation of angular position or energy selection systematically affected the data. The agreement between the experimental data and the tensor interaction curves in these experiments indicates that the tensor interaction dominates" (Rustad and Ruby 1953, p. 881). They also noted that their results were in agreement with those of Allen and Jentschke, who reported, at a meeting of the American Physical Society, that the angular correlation was $(1 + \frac{1}{3} \beta\cos\Theta)$, corresponding to the tensor interaction.[15]

In 1955, Rustad and Ruby (1955) published a much more detailed account of their experiment. This included considerable detail on the technical aspects of the apparatus, along with some data that had not been included in their preliminary account. They again showed their angular distribution for electrons with energy between 2.5 and 4.0 mc^2, this time with theoretical curves for the $S, V, T,$ and A interactions. The S and V curves, as expected, did not fit the data at all. They also presented data taken with electron energies of 4.5–5.5 mc^2 and 5.5–7.5 mc^2 (Figure 3.6). Once again, the tensor interaction was dominant. In addition, these curves provided evidence against any significant effects of gas scattering on the angular correlation. For electron energies of 4.5–5.5 mc^2, there should be a cutoff of events below an angle of 110°. For energies of 5.5–7.5 mc^2, the cutoff should be at 140°. The fact that the number of events goes to zero at these angles, as shown in Figure 3.6, indicates that scattering was not a problem. They also calculated values of α, the correlation coefficient, from their data. They found $\alpha = 0.36 \pm 0.11$ at a mean electron energy of 1.25 MeV (2.5 mc^2) and 0.31 ± 0.14 at 2.0 MeV (4 mc^2). This was compared with values $\alpha_T = \frac{1}{3}$ and $\alpha_A = -\frac{1}{3}$ predicted from pure tensor and axial vector forms, respectively. The superiority of the tensor form is again apparent.

A second series of experiments using a radioactive noble gas

---

[15] Rustad and Ruby cited an abstract, Allen and Jentschke (1952). This abstract later appeared as Allen and Jentschke (1953).

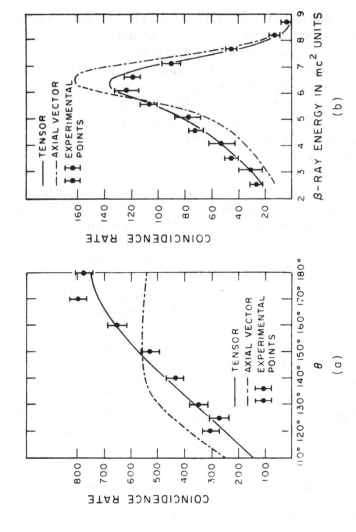

Figure 3.5. (a) Coincidence counting rate versus angle between the electron and the recoil nucleus, for electrons in the energy range 2.5–4.0 mc^2. (b) Coincidence counting rate versus electron energy for an angle of 180° between the electron and the recoil nucleus. From Rustad and Ruby (1953).

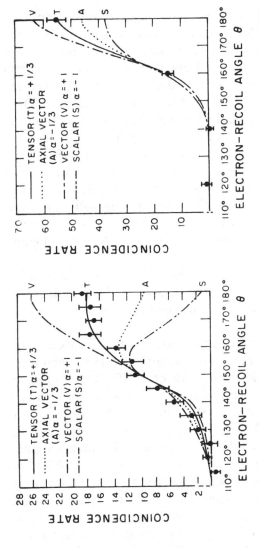

Figure 3.6. Coincidence counting rate versus angle between the electron and the recoil nucleus for (a) electrons in the energy range 4.5–5.5 mc^2 and (b) electrons in the energy range 5.5–7.5 mc^2. From Rustad and Ruby (1955).

was performed on Ne^{19}, which decayed into a positron, a neutrino, and F^{19}. The decay of Ne^{19} involved both Fermi and Gamow-Teller selection rules, so in analyzing the results to find the form of the Fermi interaction (S or V) one had to assume some form for the Gamow-Teller part of the interaction. In all of the experiments discussed in what follows the Gamow-Teller interaction was assumed to be tensor on the basis of the experimental results of Rustad and Ruby and of Allen and Jentschke.

Although the measured values of α, -0.8 ± 0.4 (Alford and Hamilton 1954, 1955), -0.21 ± 0.08 (Maxson, Allen, and Jentschke 1955), 0.14 ± 0.13 (Good and Lauer 1957), and -0.15 ± 0.2 (Alford and Hamilton 1957),[16] were not mutually consistent they all supported the "earlier conclusions that the beta interaction has the form $ST(P)$ rather than $VT(P)$" (Alford and Hamilton 1957, p. 673).

Yet another experiment consistent with the ST combination was Robson's (1955) measurement of the angular correlation in the decay of the free neutron. He obtained a value of $\alpha = +0.089 \pm 0.108$.

The agreeable consistency of the experimental results with the predictions of the ST interaction was disturbed in 1957 when Allen and his collaborators (Hermannsfeldt et al. 1957) measured the A^{35} angular correlation. This was a predominantly Fermi-type decay and was thus sensitive to the amount of S or V present in the decay interaction. They reported on two experiments. In the first, using the same apparatus they had used for their measurement on Ne^{19}, they found a value of $\alpha = 0.9 \pm 0.3$. A second experiment using only the energy spectrum of the recoil nuclei gave $\alpha = 0.7 \pm 0.17$. They pointed out that "the result that $\lambda[\alpha]$ > ⅓ in both experiments implies the existence of a vector interaction independently of any assumption about nuclear matrix elements" (Hermannsfeldt et al. 1957, p. 642; see Figure 3.7). This "rather unexpected result" (their words) caused them to carefully review the existing data on Ne^{19}. They found that the values of α were consistent either with the ST or with the VA combination of interactions. Recall that the He^6 experiments had clearly demonstrated that the Gamow-Teller part of the interac-

---

[16] Alford and Hamilton attributed the difference between this value and their earlier value to timing uncertainties.

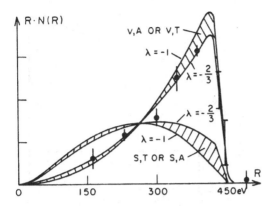

Figure 3.7. Energy spectrum of recoil ions from A^{35} decay. From Hermannsfeldt et al. (1957).

tion was tensor. "Thus, there is an apparent inconsistency between the experiments on the negatron decay of He^6 and the positron decays of Ne^{19} and A^{35}" (Hermannsfeldt et al. 1957, p. 643).

Further confusion was added when Cavanagh (1958a) reported Ridley's results on Ne^{23} at the Rehovoth Conference in September 1957.[17] Ne^{23} was believed to be a substantially pure Gamow-Teller transition and theory would predict $\alpha = +\frac{1}{3}$ for a pure tensor interaction and $\alpha = -\frac{1}{3}$ for a pure axial vector. Ridley found a value of $\alpha = -0.05 \pm 0.10$, in serious disagreement with the Rustad-Ruby result.

At this same conference, Konopinski (1958) summarized the situation using a Scott diagram that plotted the value of α against x, which was essentially a measure of the relative amount of Gamow-Teller and Fermi interactions present in a particular decay. At $x = 0$, or a pure Gamow-Teller transition, $\alpha(0) = \alpha_{GT} = \frac{1}{3}[G_T^2 - G_A^2]/[G_T^2 + G_A^2]$. This gives $\alpha = +\frac{1}{3}$ for pure T and $\alpha = -\frac{1}{3}$ for pure A, as noted previously. For $x = 1$, or a pure Fermi transition, $\alpha(1) = \alpha_F = [G_V^2 - G_S^2]/[G_V^2 + G_S^2]$. Thus, $\alpha = +1$ for pure V, and -1 for pure S. For a superposition of Fermi and Gamow-Teller interactions $\alpha(x) = \alpha_F x + \alpha_{GT}(1 - x)$. Konopinski's summary along with the theoretical predictions are shown in Figure 3.8. Konopinski summarized the situation as

[17] Cavanagh reported the results in B. Ridley (1954).

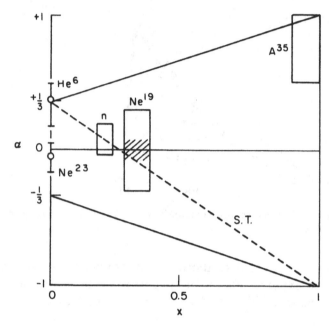

Figure 3.8. Electron–neutrino correlation coefficient versus Fermi fraction. From Konopinski (1958).

follows. The experiments on He⁶, n, and Ne¹⁹ were consistent with *ST*, whereas the experiments on Ne²³, n, Ne¹⁹, and A³⁵ were consistent with *VA*. "It seems too early to choose between them" (Konopinski 1958, p. 330).

The evidence from muon decay, discussed earlier, was consistent with *ST*, although no one seems to have attempted to fit a *VA* combination.

4

The discovery of parity nonconservation[1]

In late 1956 and early 1957 the situation changed dramatically. Following a suggestion by Lee and Yang (1956) that parity, or mirror symmetry, might be violated in the weak interactions, which included β decay, a series of experiments by Wu and her collaborators (1957), by Garwin, Lederman, and Weinrich (1957), and by Friedman and Telegdi (1957a) showed conclusively that this was the case. This discovery had serious implications for the previous analyses of β decay, suggested new experiments, and pointed the way toward a new theory of β decay.

We can summarize the history of this discovery as follows. During the 1950s the physics community was faced with what was known as the "θ–τ puzzle." On one set of accepted criteria, that of identical masses and lifetimes, the θ and τ particles appeared to be the same particle. On another set of accepted criteria, that of spin and parity, they appeared to be different. The spin and parity analysis was performed on the decay products, two pions for the θ and three pions for the τ. Parity conservation was assumed in these decays and the spin and parity of the θ and τ were inferred. There were several attempts to solve this puzzle within the framework of currently accepted theories, but all of these were unsuccessful.

In 1956, Lee and Yang recognized that a possible solution to the problem would be the nonconservation of parity in the weak interactions. If parity were not conserved then the θ and τ would merely be two different decay modes of the same particle. This led them to examine the existing evidence in favor of parity conservation. They found, to their surprise, that although earlier

---

[1] For a detailed history of the discovery of parity nonconservation see Franklin (1986, ch. 1).

experiments supported the conservation of parity in the strong and electromagnetic interactions to a high degree of accuracy, there was, in fact, no evidence in favor of parity conservation in the weak interactions.

Lee and Yang suggested several possible tests in their paper. The abstract is a masterpiece of understatement. "The question of parity conservation in β decays and in hyperon and meson decays is examined. Possible experiments are suggested which might test parity conservation in these interactions" (Lee and Yang 1956, p. 254). Although Lee and Yang suggested several possible experiments I shall concentrate on only two, the β decay of oriented nuclei, and the sequential decay $\pi \to \mu \to e$, because these were the first experiments done and provided the crucial evidence for the physics community.

Lee and Yang described these experiments as follows:

A relatively simple possibility is to measure the angular distribution of the electrons coming from the β decays of oriented nuclei. If θ is the angle between the orientation of the parent nucleus and the momentum of the electron, an asymmetry of distribution between θ and 180°–θ constitutes an unequivocal proof that parity is not conserved in β decay (Lee and Yang 1956, p. 255).

In the decay processes

$$\pi \to \mu + \nu, \tag{5}$$
$$\mu \to e + \nu + \nu, \tag{6}$$

starting from a π meson at rest, one could study the distribution of the angle θ between the μ-meson momentum and the electron momentum, the latter being in the center of mass of the μ meson. If parity is conserved in neither (5) nor (6), the distribution will not in general be identical for θ and $\pi-\theta$ [180°–θ]. To understand this, consider first the orientation of the muon spin. If (5) violates parity conservation, the muon would be in general polarized in its direction of motion. In the subsequent decay (6), the angular distribution problem with respect to θ is therefore closely similar to the angular distribution problem of β rays from oriented nuclei, which we have discussed before (Lee and Yang 1956, p. 257).

The first experiment was performed by Wu and her collaborators (1957). It consisted of a layer of polarized Co^{60} nuclei and a single electron counter, which was located either parallel to or

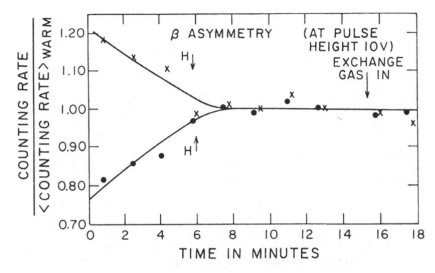

Figure 4.1. Relative counting rates for β particles from the decay of oriented Co^{60} nuclei for different nuclear orientations (field directions). From Wu et al (1957).

antiparallel to the orientation of the nuclei. The direction of the polarization of the nuclei could be changed and any difference in counting rate in the fixed electron counter observed. Their results are shown in Figure 4.1. With the counter antiparallel to the nuclear polarization the ratio of counts observed when the nuclei were polarized to when they were not was 1.20. With the counter parallel to the polarization the ratio was 0.80. This was a clear asymmetry. They concluded, "If an asymmetry in the distribution between θ and 180°–θ (where θ is the angle between the orientation of the parent nuclei and the momentum of the electrons) is observed, it provides unequivocal proof that parity is not conserved in β decay. This asymmetry has been observed in the case of oriented Co^{60}" (Wu et al. 1957, p. 1413).

The second experiment was performed using two different techniques by Garwin, Lederman, and Weinrich (1957) and by Friedman and Telegdi (1957a). The Garwin experiment consisted of stopping μ mesons from π-meson decay in a block of carbon and detecting the electrons from the subsequent decay of the μ meson. Rather than move the electron counter to observe the electron asymmetry in the decay angular distribution, they found it easier to fix the electron counter and to precess the spins of the μ mesons.

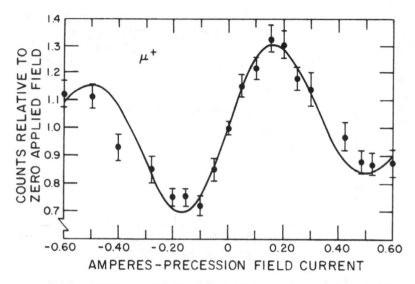

Figure 4.2. Relative counting rate as a function of precession current. If parity
were conserved the curve would be a horizontal line. From Garwin et al. (1957).

They found a sinusoidal variation in counting rate, as opposed to
the flat distribution expected if parity were conserved (see Figure
4.2). Their statistically overwhelming effect (22 standard deviations) led them to conclude that parity was not conserved. More
informally, Lederman called Lee at 7 AM and announced, "Parity
is dead" (Lee 1971, p. 12).

Friedman and Telegdi performed the same experiment using a
different technique. They stopped π^+ mesons in a nuclear emulsion, in which the μ^+ from the resulting decay also stopped. They
looked at the distribution of decay electrons relative to the muon
direction. They found a backward–forward asymmetry of 0.062 ±
0.027. A note added in proof changed this result to 0.091 ± 0.022.
They, too, concluded that parity was not conserved.

This was also the conclusion of the physics community. It is
fair to say that as soon as any physicist saw the experimental
results they were convinced that parity was not conserved in the
weak interactions.

This discovery had implications for the earlier analyses of β
decay, which had assumed that parity was conserved. Lee and
Yang had pointed out that the earlier experiments had not measured any quantity that might have demonstrated parity noncon-

servation. The new discovery did, however, complicate the analysis of the old results and modified some of the earlier conclusions. Konopinski summarized the situation in his review talk, "Theory of the Classical β-Decay Measurements" (Konopinski 1958). "The 'classical' measurements had led us to form certain conclusions. Those conclusions were made under then unquestioned restrictions [including parity conservation] which, now, must be discarded. This is just what makes a review of the old interpretations imperative." He also noted, however, " . . . that the actual data produced in the classical lines of investigation are hard facts. They cannot be ignored when interpreting the new experiments" (Konopinski 1958, p. 320).

One of the most important changes required was in the form of the interaction. We may write the interaction Hamiltonian for β decay as

$$H = g \sum_x C_x \, (P_x \cdot \rho_x) + \text{complex conjugate},$$

where the P_x are the S, V, T, A, and P forms of the interaction containing the proton and neutron wavefunctions and the ρ_x are the expressions containing the electron and neutrino wavefunctions. As Konopinski pointed out, one could not really say which of the pair ρ_s or ρ_p was the scalar and which the pseudoscalar. Another perfectly reasonable Hamiltonian would be

$$H' = g \, [C'_S \, P_S \, \rho_P + C'_P \, P_P \, \rho_S + \dots].$$

Each of these Hamiltonians gives the same predicted experimental results. If both are included, however, interferences are produced that violate parity conservation. This had been previously excluded. With the discovery of parity violation the sum of the two Hamiltonians, or something equivalent, was required. This gave ten coupling constants to be determined experimentally rather than the five required previously.

A second doubling of the number of coupling constants was also required if one considered that time reversal invariance (another unquestioned symmetry principle) was violated. This allowed each of the coupling constants C_x, C'_x to be complex. The discovery that parity conservation, which had also been previously unquestioned, was violated made this suggestion more plausible. Pauli (1957) suggested that yet another doubling of the coupling constants was possible if lepton conservation was violated. The number of leptons or light particles (muons, electrons, and neutrinos) was thought to be conserved in particle interactions. Thus, the neutron decayed into a proton, an electron, and an antineutrino. There are no leptons in the initial state, and the electron and antineutrino combine to give a lepton number equal to 0. Pauli discussed the possibility that there might also be a decay of the neutron in which a neutrino was emitted, and which would violate lepton conservation.

In general, theoretical physicists considered only the effect of parity violation in their reanalyses of weak interactions. These new considerations changed some of the conclusions of the previous analyses of β decay. The results still required the presence of both Fermi- and Gamow-Teller-type transitions. In addition, there would be no interferences between these transitions because one averaged over nuclear orientations and summed over electron polarizations. The new analysis still allowed the determination of the proportion of Fermi and Gamow-Teller components of the radiation and whether it was the S or V interaction responsible for the Fermi transition or T or A for the Gamow-Teller type. Although an expression for the imaginary part of the coupling constants appeared in the angular correlation coefficient, the experiments were not sensitive enough, by a factor of approximately 20, to detect such effects.

The conclusions to be drawn from the absence of Fierz interference terms had to be modified. Previously, the measurement of Sherr and Miller (1954) on the ratio of K-capture to positron emission in Na^{22} had determined either C_A/C_T or C_T/C_A to be equal to -1 ± 2 percent. In the new analysis, all one could say was that $b_{GT} = \text{Re}[C_T C_A{}^* + C'_T(C'_A)^*]/[|C_T|^2 + |C_A|^2 + |C'_A|^2 + |C'_T|^2] = -0.01 \pm 0.02$ for the Gamow-Teller interaction. A similar analysis by Sherr and Gerhart (1956) set $b_F = 0.00 \pm 0.10$ for the Fermi interaction. This was the chief effect of the new considerations on the old interpretations.

Even before the experimental results demonstrating the nonconservation of parity were known,[2] theoretical physicists attempted to incorporate parity violation into the theory of weak interactions in a natural, plausible way. This was done indepen-

---

[2] Lee and Yang did know of the preliminary results of Wu's experiment.

dently by Lee and Yang (1957), by Landau (1957), and by Salam (1957). This involved, in all three cases, a two-component theory of the neutrino. In conventional relativistic physics the neutrino wavefunction contained four components: two for a neutrino with spin either parallel to or antiparallel to its momentum and two for an antineutrino with the same properties. In the two-component theory, the neutrino had only a spin parallel to its momentum and the antineutrino, antiparallel, or vice versa.

This theory clearly violated parity conservation. We may see this as follows. A three-dimensional space reflection reverses all of the components of the momentum so that the momentum vector is reversed. The spin vector, however, remains unchanged. Thus, the mirror image differs from the original and parity is not conserved. As Lee and Yang noted, this fact had been known for some time and Pauli had rejected such a theory, originally proposed by Weyl (1929), precisely because it violated parity conservation. "However, as the derivation shows, these wave equations are not invariant under reflections (interchanging left and right) and thus are not applicable to physical reality" (Pauli 1933, p. 226). The two-component theory also required that the mass of the neutrino be identically zero and that charge conjugation invariance (particle–antiparticle symmetry) be violated.

There were several important experimental implications of this theory. It predicted both the asymmetry in the β decay of oriented nuclei and that the electrons emitted in the decay would have a polarization equal to v/c, where v is the velocity of the electron and c is the speed of light.

Perhaps most important for our discussion was the analysis of muon decay. All three papers reached the same conclusion. They considered three possibilities for muon decay:

1. $\mu \rightarrow e + \nu + \bar{\nu}$,
2. $\mu \rightarrow e + 2\nu$,
3. $\mu \rightarrow e + 2\bar{\nu}$,
   where ν = neutrino and $\bar{\nu}$ = antineutrino.

In case (1) the two-component theory required that the S, T, and P couplings equal zero and that the Michel parameter in the decay spectrum $\rho = \frac{3}{4}$. For cases (2) and (3) $\rho = 0$. This was inconsistent with the experimental results of Sargent et al. (1955) of $\rho = 0.64 \pm 0.10$ and of Bonetti et al. (1956) of $\rho = 0.57 \pm 0.14$, a point noted by both Landau and by Lee and Yang. Thus, the decay interaction for muons had to be a combination of the vector and axial vector (V and A) forms.

During the next few months, the nonconservation of parity received additional confirmation from experiments on other weak decay processes and from repetitions of the three original experiments. The asymmetry in π–μ–e decay was observed in an experiment using bubble chambers, yet another technique (Abashian et al. 1957). Other parity-violating effects such as the circular polarization of γ rays emitted after β decay (Schopper 1957; Boehm and Wapstra 1957), the circular polarization of bremsstrahlung emitted by β rays (Goldhaber, Grodzins, and Sunyar 1957), and the longitudinal polarization of positrons emitted in radioactive decay (Hanna and Preston 1957; Page and Heinberg 1957) were also seen. The longitudinal polarization of electrons was further confirmed for Co^{60} and for other elements (De Waard and Poppema 1957; Frauenfelder et al. 1957a). Both Wu and her collaborators (Ambler et al. 1957) and Friedman and Telegdi (1957b) presented further results and analysis.

The evidence supporting the two-component theory of the neutrino was not as clear as that in favor of parity violation. All of the experiments discussed above supported the theory in a qualitative way. With regard to the quantitative prediction that the polarization of the emitted electrons (positrons) should equal v/c, the evidence was ambiguous. In general, for pure Gamow-Teller transitions there was no discrepancy. For mixed Fermi and Gamow-Teller transitions or for pure Fermi decays the prediction was not confirmed.

De Waard and Poppema (1957) detected the polarization of decay electrons emitted from Co^{60} and P^{32} by Mott scattering from nuclei. They found polarizations of -0.49 ± 0.11 and -0.50 ± 0.11 from Co^{60} and P^{32}, respectively. This was slightly lower than the theoretically predicted value of 0.66. They noted, however, that depolarization effects in the scattering foils might account for the discrepancy.

There was, however, a strong preponderance of evidence supporting the two-component neutrino from the study of Co^{60} decay, a pure Gamow-Teller transition. Schopper (1957) and Boehm and Wapstra (1957) measured the polarization of γ rays emitted after the β decay and found that α, the asymmetry coefficient, was -0.41 ± 0.07 and -0.40 ± 0.09, respectively, in good agreement with the theoretically predicted value of $-\frac{1}{3}$. Further support came from the Co^{60} experiments of Wu (Ambler et al. 1957), Frauenfelder (1957a), and Cavanagh (1957). Schopper also found $\alpha = +0.39 \pm 0.08$ for Na^{22}, in agreement with the theoretical value of $+\frac{1}{3}$. The change in the sign of α was also predicted by theory. The results of De Waard and Poppema were also contradicted by those of Frauenfelder et al. (1957b), who found that for both P^{32} and Pr^{144} the polarizations were equal to v/c.

Evidence from mixed Fermi and Gamow-Teller transitions did not agree with the two-component theory. Ambler et al. (1957) found for Co^{58}, a positron emitter, that the sign of the asymmetry changed as expected, but that the polarization effect was only one-third that of Co^{60}. They concluded that this result could not be explained if their theoretical assumptions included (1) the two-component neutrino, (2) the dominance of S and T interactions in β decay, and (3) time reversal invariance. This discrepancy was further supported by the work of Frauenfelder and his collaborators (1957c). They investigated the mixed transitions of Sc^{46} and Au^{198}. For Sc^{46} $v/c = 0.6$ whereas the measured polarization was -0.34 ± 0.10. For Au^{198} they found, for different methods:

v/c	Polarization
0.55	-0.05 ± 0.06
0.6	-0.06 ± 0.05
>0.78	$+0.05 \pm 0.12$
0.78–0.92	$+0.02 \pm 0.23$

These results were in clear disagreement with the theoretical predictions that the polarization equal v/c. They discussed the earlier difficulties of Wu and her collaborators on Co^{58} and pointed out that the Au^{198} and Sc^{46} results could be explained if one assumed that the Fermi transitions occurred through the vector interaction. They noted that this was in agreement with the then recent result on the angular correlation in A^{35} by Hermannsfeldt et al. (1957). They also discussed the ambiguity of the evidence from angular correlation experiments mentioned earlier, and that the results of Boehm and Wapstra seemed to favor S as the Fermi interaction. They noted that the evidence for pure Gamow-Teller transitions favored the two-component neutrino, though that for mixed transitions in Au^{198}, Sc^{46}, and Co^{58} did not. They mentioned a suggestion made by Alder, Stech, and Winther[3] that perhaps parity was conserved in Fermi transitions and not in Gamow-Teller transitions. To test this, they measured the polarization of positrons emitted in Ga^{66} decay, which was believed to be a pure Fermi transition [Frauenfelder et al. 1957d). They found that the polarization was $+0.09 \pm 0.31$, whereas v/c was 0.98. This supported the suggestion of parity conservation in Fermi transitions.

The situation with regard to a Universal Fermi Interaction, that would apply to all weak interactions, was unclear. As Lee remarked at the Rochester Conference (April 15–19, 1957),

Beta decay information tells us that the interaction between (p,n) and (e,ν) is scalar and tensor, while the two-component neutrino theory plus the law of conservation of leptons implies that the coupling between (e,ν) and (μ,ν) is vector. This means that the Universal Fermi Interaction cannot be realized in the way we have expressed it. If all these coupling types turn out to be experimentally correct, we prefer to think that the similarity in coupling constants cannot be accounted for in terms of such a limited scheme. Rather it is a universal feature of all weak interactions, and not just those involving leptons. Nevertheless, at this moment it is very desirable to recheck even the old beta interactions to see whether the coupling is really scalar, a point we shall return to later (Lee 1957, p. VII-7).

Lee did not question the conclusion that the Gamow-Teller interaction was tensor. The evidence of the polarization of electrons from mixed transitions and the electron–neutrino angular correlation experiments led one to question the conclusion that S and T were the dominant interactions and that V was excluded. The inclusion of V, along with T, would result in neutrinos of opposite helicities being emitted in Fermi and Gamow-Teller transitions, a prospect that did not appeal to theorists.[4] Another attempt to resolve these difficulties was the twin-neutrino hypothesis proposed by Mayer and Telegdi (1957). In this model there were two different two-component neutrinos, with opposite screw sense or helicity, along with their respective antiparticles.

[3] Their citation was a private communication and an unpublished University of Illinois report.

[4] At this time, Schwinger (1957) did offer a VT combination based on his theory of intermediate bosons.

One of these was associated with the V and T interactions, and the other with S and A. The authors pointed out that such a model could explain the discrepancy from the Co^{58} results.

During the summer of 1957, the situation became clearer. At the Rehovoth Conference (Sept. 8–14, 1957), experiments that demonstrated electron polarization equal to v/c,[5] circular polarization of bremsstrahlung produced by decay electrons (Cohen et al. 1958), β–γ circular polarization correlation (Schopper 1958 and Steffen 1958), and β-decay asymmetries (Postma et al. 1958) were reported. These confirmed the two-component theory of the neutrino. Other reports removed some of the discrepancies for the theory. De Waard, Poppema, and Van Klinken (1958) reported remeasurements of the polarization from Co^{60}, P^{32}, Tm^{170}, and Au^{198}. They withdrew their earlier results on electron polarization because of experimental problems in the positioning of the source and the scattering foils and from depolarization and scattering effects due to the thickness of the source and the scattering foils. In particular, they withdrew their previous result for Au^{198} because of bad source geometry and reported new results in agreement with those for P^{32} and the two-component theory. They also found that for 170 KeV electrons from Sc^{46} and Co^{60} the polarization was 1.00 ± 0.015, again in agreement with theory. Wu (1958) also reported results on Au^{198} that contradicted the earlier values. She found polarizations of $(-1.02 \pm 0.19)\, v/c$ and $(-0.95 \pm 0.25)\, v/c$, also as expected.

Evidence was also presented that parity was not conserved in Fermi transitions. Previously the lack of polarization of positrons from Ga^{66} decay had argued for parity conservation. Frankel et al. (1958) measured the polarization of these positrons by their annihilation in polarized matter and found a polarization equal to v/c. A similar conclusion was reached by Deutsch et al. (1957, paper received August 2, 1957), who concluded that the positrons from both Ga^{66} and Cl^{34} decay, another Fermi transition, had high polarization, as expected.

During this summer the results of Boehm and Wapstra (1958, published on January 15, 1958) also became known.[6] They mea-

---

[5] Bielein, Fleischmann, and Wegener (1958), Cavanagh (1958b), Frauenfelder (1958), and Lipkin et al. (1958).
[6] Sudarshan and Marshak (1958).

sured the β–γ circular polarization correlation for Na^{24}, Sc^{44}, Sc^{46}, V^{48}, and Co^{58}. They concluded,

This work, particularly the experiment on Sc^{46}, establishes the presence of a large interference between Gamow-Teller and Fermi couplings. It therefore excludes a pure V,T and a pure S,A interaction. It also disproves the validity of the twin-neutrino theory [Boehm and Wapstra 1958, p. 461]. All results [of this experiment] are in agreement with the 2-component neutrino theory and the assumption of V,A or alternatively S,T,P interaction (p. 456).

The situation at the end of the summer of 1957 was as follows. Parity nonconservation had been conclusively demonstrated and there was strong experimental support for the two-component theory of the neutrino. That theory plus the conservation of leptons led to the conclusion that the weak interaction responsible for muon decay had to be a VA combination. Although most of the experimental evidence from nuclear β decay was consistent with a VA interaction, the seemingly conclusive evidence from the He^6 angular correlation experiment of Rustad and Ruby (1953, 1955) gave T as the interaction. The failure to observe the decay of a pion into an electron plus a neutrino also argued against a VA interaction.

5

The *V-A* theory of weak interactions
and its acceptance

5.1 THE SUGGESTION OF *V-A* THEORY

The situation discussed at the end of the last chapter was described in papers that proposed that a Universal Fermi Interaction, one that applied to all weak interactions, was a linear combination of *V* and *A*. The theory was offered by Sudarshan and Marshak (1957, 1958) and by Feynman and Gell-Mann (1958). This was exactly the opposite conclusion drawn four years earlier by Konopinski and Langer, who had stated, "As we shall interpret the evidence here, the correct law must be what is known as an *STP* combination" (1953, p. 261).

Sudarshan and Marshak examined the available evidence from nuclear β decay and other weak interactions, including strange particle decays, and concluded that the only possible choice for a Universal Fermi Interaction was a linear combination of *V* and *A,* even though there was evidence apparently in conflict with this choice. The four experiments cited in opposition to the *V-A* theory were

1. The electron–neutrino angular correlation experiment on He^6 by Rustad and Ruby (1953, 1955), which gave *T* as the β-decay interaction.
2. The sign of the electron polarization from muon decay.[1]
3. The frequency of the electron mode in pion decay.[2]

---

[1] Sudarshan and Marshak cited L. Lederman's private communication to P. T. Matthews.
[2] Anderson and Lattes had set an upper limit of 10^{-5} for this decay. Sudarshan and Marshak cited a private communication from Gell-Mann. This result was later published as Anderson and Lattes (1957).

4. The asymmetry in polarized neutron decay, which was smaller than predicted.[3]

They suggested,

All of these experiments should be redone, particularly since some of them contradict the results of other recent experiments on the weak interactions. If any of the above four experiments stands, it will be necessary to abandon the hypothesis of a universal $V + A$[4] four fermion interaction or either or both of the assumptions of a two-component neutrino and/or the Law of Conservation of Leptons (Sudarshan and Marshak 1957, pp. 126–7).

They also pointed out that the theory had some very attractive theoretical features. It provided a natural mechanism for parity violation in strange particle decays leading to the θ- and τ-decay modes of the K meson. It behaved in the same way for both scalar and pseudoscalar mesons which allowed it to treat pions and kaons in the same way. It had the added feature that the neutrino emitted in both Fermi and Gamow-Teller transitions had the same handedness (right or left), which was not true for either the VT or SA combinations.

This theoretical elegance was also emphasized by Feynman and Gell-Mann.

It is amusing that this interaction satisfies simultaneously almost all the principles that have been proposed on simple theoretical grounds to limit the possible β couplings. It is universal, it is symmetric, it produces two-component neutrinos [and thus violates parity], it conserves leptons, it preserves invariance under CP and T . . . (Feynman and Gell-Mann 1958, pp. 197–8).

In fact, the theoretical arguments led Feynman and Gell-Mann to a stronger statement about the He^6 angular correlation experiment. "These theoretical arguments seem to the authors to be strong enough to suggest that the disagreement with the He^6 recoil experiment and with some other less accurate experiments indicates that these experiments are *wrong* [emphasis added]. The $\pi \rightarrow e + \nu$ may have a more subtle solution" (Feynman and Gell-Mann 1958, p. 198).

This is not to say that Feynman and Gell-Mann did not care

---

[3] Sudarshan and Marshak cited Novey and Telegdi's private communication from Dr. Stech. See also Burgy et al. (1957, 1958).
[4] The relative phase of the V and A amplitudes was not yet fixed.

about the experimental evidence, but only that theoretical elegance was an important consideration for them.

After all the theory also has had a number of successes. It yields the rate of μ decay to 2%[5] and the asymmetry in the direction in the $\pi \rightarrow \mu \rightarrow$ e decay chain. For β decay, it agrees with the recoil experiments in A^{35} indicating a vector coupling, the absence of Fierz interference terms distorting the allowed spectra, and the more recent electron spin polarization measurements in β decay (Feynman and Gell-Mann 1958, p. 198).

One point that should be emphasized here is that the interactions among the community of physicists consist of far more than published papers. The informal network of conversations, preprints, and conference papers provides an extremely important source of information, particularly in a rapidly developing field such as weak interaction physics in 1957. We also see that theorists and experimentalists are quite aware of each other's work and the fruitful interaction of theory and experiment. Thus, in the paper by Sudarshan and Marshak, the important experimental results on Ga^{66} decay (private communication from Dr. Stech); electron polarization in Au^{198} (Frauenfelder et al., to be published); electron polarization in Co^{58} (C. S. Wu, *Proceedings of the Seventh Rochester Conference 1957,* which was not yet published); electron polarization in Sc^{46} (Wapstra and Boehm, to be published); the asymmetry in polarized neutron decay (Novey and Telegdi; private communication from Dr. Stech); Anderson's and Lattes's upper limit of 10^{-5} for the electron decay of the pion (private communication from Prof. Gell-Mann); the scalar nature of the K meson (private communication from R. L. Walker and R. R. Wilson); radiative corrections to the ρ parameter (Kinoshita and Sirlin, to be published); and the sign of the electron polarization in muon decay (L. Lederman, private communication to P. T. Matthews) were all transmitted this way. The paper by Sudarshan and Marshak was itself given at a conference, with a short formal publication later. Feynman and Gell-Mann also acknowledged conversations with Boehm, Wapstra, Stech, Marshak, and Sudarshan.

---

[5] They calculated a lifetime for the muon of $(2.26 \pm 0.04) \times 10^{-6}$ s compared to a measured value of $(2.22 \pm 0.02) \times 10^{-6}$ s.

5.2 THE RESOLUTION OF THE DISCREPANCIES AND THE CONFIRMATION OF THE *V-A* THEORY

During late 1957 and through 1958, the four experiments that had been noted as anomalies for the *V-A* theory were repeated. The new results were in agreement with that theory, the discrepancies seen as resolved, and the theory accepted. One might worry that the theoretical presuppositions of the experimenters had an effect on the results, but the history does not support this. This issue will be discussed in detail subsequently. One should remember, however, that it was the Rustad-Ruby experiment that strongly supported the tensor interaction, and not vice versa. Before that experiment the question of the nature of the Gamow-Teller interaction was unanswered.

5.2a The angular correlation in He^6

Perhaps the most important piece of evidence favoring tensor (T) as the Gamow-Teller part of the weak interaction had come from the angular correlation experiment on He^6 by Rustad and Ruby (1953, 1955). Their data (see Figures 3.5 and 3.6) clearly favored the tensor interaction over the axial vector interaction. Following the proposal of the *V-A* theory this experiment was carefully reexamined by Wu and Schwarzschild (1958) and by Rustad and Ruby themselves. The reanalysis by Wu and Schwarzschild was almost unprecedented in physics and will be discussed in detail.

One of the important factors in the calculation of the Rustad-Ruby result was the assumption that all of the He^6 decays came from the source volume. If, however, there had been a significant amount of gas in the collimating chimney below the source volume (see Figure 3.4) then both their measured correlation and their conclusion of tensor dominance might have had to be modified. The argument, given by Wu and Schwarzschild, was as follows. First, the efficiency of the β counter changes considerably with the angle between the counter and the recoil detector. For decays from the chimney at 180° the electrons do not pass through much material and a large number pass unimpeded through the holes in the diaphragm. At 90°, only those electrons with sufficient energy to penetrate the pumping diaphragm will be detected. If one combines this with the fact that the geometrical efficiency, or solid angle, of the recoil detector is much larger for decays in the chimney than for decays in the source volume, then one finds a considerable change in the total detection efficiency. The question to be answered then was what was the gas density in the chimney. Wu and Schwarzschild performed an approximate calculation and found that the gas pressure at the lower end of the first diaphragm had to be greater than 12 percent of that in the source volume.

A better estimate was obtained by constructing a physical analogue of the gas system. Because the gas pressure in the system was so low, one could neglect collisions between the molecules and consider only reflections from the walls of the apparatus in computing the pressure gradient. This is similar to optical diffusion from a perfectly reflecting surface. Wu and Schwarzschild constructed a scale model ten times larger than the actual Rustad-Ruby apparatus. They made the inner walls of both the source volume and the collimating chimney of the model highly reflecting by coating them with MgO. They placed a diffuse light source in the arm of the source volume to simulate the entering gas. They then measured the relative light intensity at various parts of the model as an indication of the relative gas pressure. They found that the light intensity was constant inside the source volume and then dropped linearly through the chimney. They combined this with the solid angle subtended by the recoil detector and the effective volume of gas and concluded, "the decay due to the He^6 gas in the chimney is not insignificant in the correlation results" (Wu and Schwarzschild 1958, p. 6).

Another important factor in estimating the correction to the correlation results was the fact that the pumping diaphragm was semitransparent to the decay electrons. Those electrons suffered energy loss and multiple scattering that might have destroyed their correlation relations. This was estimated by measuring the beta spectrum of Rh^{106} with a polystyrene diaphragm and with both the diaphragm and a brass plate. In addition, backscattering of electrons from the diaphragm walls was considerable. They measured the Rh^{106} spectrum with the source at the center of the source volume and at the mouth of the chimney. They found an excess of low-energy electrons when the source was in the chimney, which they attributed to backscattering.

A precise correction calculation was "obviously impossible"

(their words). Wu and Schwarzschild calculated the correction for two extreme cases. The first included the change in overall efficiency using only the polystyrene diaphragm results. The second calculation added the assumption that the electrons penetrating an opaque diaphragm maintained their correlation relations while the excess due to the semitransparency of the diaphragm had lost their correlation and had an isotropic correlation. They regarded the second calculation as more realistic. The results of these two calculations along with the original Rustad-Ruby results are given in Figure 5.1 and "are more in favor of axial vector than tensor contradictory to the original conclusion" (Wu and Schwarzschild 1958, p. 8). The work of Wu and Schwarzschild obviously cast doubt on the original Rustad-Ruby conclusion, although it could not be used to support the view that axial vector, A, was the form of the interaction.

That work was completed in November 1957,[6] and was known to Rustad and Ruby, who were doing their own reexamination. No record of this reexamination seems to be available[7] but in a postdeadline paper given at the January 1958 meeting of the American Physical Society, Rustad and Ruby suggested that their earlier result might be wrong.[8]

The He[6] angular correlation experiment was redone by Allen and his collaborators (Hermannsfeldt et al. 1958). Their results were quite different from those of Rustad and Ruby, but consistent with Allen's earlier work (Allen, Paneth, and Morrish 1949). The angular correlation coefficient α (λ in the paper) should be $+\frac{1}{3}$ for the tensor interaction and $-\frac{1}{3}$ for the axial vector interaction. Their results are shown in Figure 5.2. They found $\alpha = -0.39 \pm 0.02$, clearly favoring axial vector. The excess number of recoil ions above the calculated curve below 700 eV was attributed to multiply charged recoil ions which were also detected by the apparatus. Those points were not used in the analysis. They were concerned that their result differed from the predicted value by more than the statistical error. They suggested that the

[6] Letter from C. S. Wu to P. Cavanagh, May 19, 1958. I am grateful to Brian Ridley for providing me with this letter.
[7] Dr. S. Ruby, private communication.
[8] There are no abstracts of postdeadline papers, but the talk is cited in Burgy et al. (1958) and Commins and Kusch (1958). Dr. Ruby (private communication) remembers the tone of the paper as mea culpa.

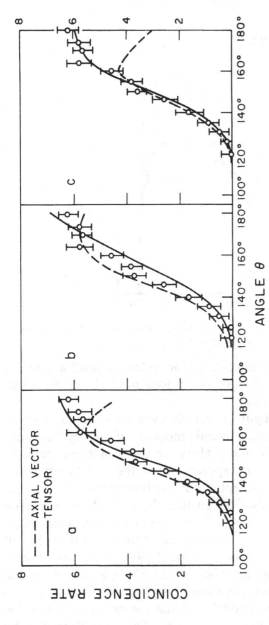

Figure 5.1. Angular correlation data of Rustad and Ruby (1953, 1955) as corrected by Wu and Schwarzschild (1958). (a) Semitransparent diaphragm. (b) Opaque plus semitransparent diaphragm. (c) Original data of Rustad and Ruby. From Wu and Schwarzschild (1958).

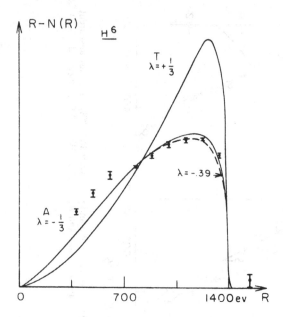

Figure 5.2. Distribution of recoil ions from the decay of He^6 as a function of the energy R. From Hermannsfeldt et al. (1958).

linearity of the spectrometer–detector system, a small amount of impurity, or multiply charged recoil ions might be responsible, but came to no conclusion.

An interesting sidelight to this discussion was an experiment to measure the spin and magnetic moment of He^6 by Commins and Kusch (1958). If the spin of He^6 were one then the discrepancy between the Rustad-Ruby experiment and the V-A theory could be resolved. Although accepted nuclear theory and previous experimental work had established that the spin of an even-even nucleus (He^6 was one) was zero, the point had not been tested for He^6. They found a value for the magnetic moment of He^6 of 0.09 nuclear magnetons, with an upper limit of 0.16 nuclear magnetons. They could find no model of the He^6 nucleus which had a spin of one and would give a magnetic moment as small as 0.16 nuclear magnetons. They concluded that the most probable value of the spin of the He^6 nucleus was zero. Thus, the Rustad-Ruby anomaly could not be solved by an incorrect spin for the He^6 nucleus.

Commins and Kusch did not, in fact, expect to find a spin of one. Their cited references indicate that they were aware of the

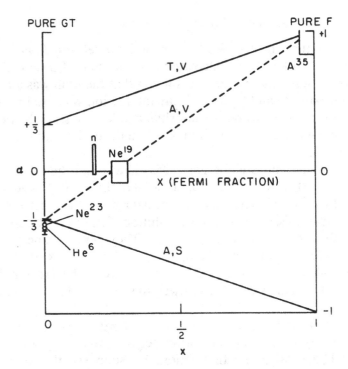

Figure 5.3. Konopinski's (1959) summary of the angular correlation results.

Rustad-Ruby modification of their conclusion and of Allen's work. This, along with the other evidence on nuclear structure, made such a result unlikely.[9] Nevertheless the anomaly for the *V-A* theory, which had so much other confirmation and theoretical attractiveness, made even an unlikely search worthwhile.

By early 1959 all of the evidence from the angular correlation experiments was in agreement with the *V-A* theory. In his review article, which surveyed the literature through March 1, 1959, Konopinski (1959) presented a figure (Figure 5.3) summarizing the situation. He included the most recent work of Allen and his collaborators (Hermannsfeldt et al. 1959). Their experiments on He^6, Ne^{19}, Ne^{23}, and A^{35} concluded, "the experimental results are consistent if we assume that the dominant beta-decay interaction is *VA*."

---

[9] I was an undergraduate assistant on this experiment and participated in some of the discussions.

5.2b The electron decay of the pion

We recall that in 1949 Ruderman and Finkelstein (1949) and Steinberger (1949b) had calculated that the ratio of $\pi-e/\pi-\mu$ decays was approximately 10^{-4}, assuming that the pion was a pseudoscalar particle and that the decay interaction was axial vector. However, as noted by both Sudarshan and Marshak and by Feynman and Gell-Mann, the best measurements of that ratio seemed to be too low.

Lokanathan and Steinberger (1955) had found the ratio to be $(-0.3 \pm 0.9) \times 10^{-4}$ with an upper limit of 0.6×10^{-4}. Anderson and Lattes (1957), using a double-focusing magnetic spectrometer of high transmission and good resolution, found no evidence at all for the electron decay of the pion. They gave a value for the ratio of electron to muon decays of $(-0.4 \pm 9.0) \times 10^{-6}$, and a probability of 1 percent that the value would be greater than 2.1×10^{-5}. This was clearly lower than the *V-A* prediction of 10^{-4}.

Anderson and Lattes pointed out that Morpurgo (1957) had found a way to forbid the electron decay of the pion, consistent with a Universal Fermi Interaction. He supposed that only the scalar and tensor (possibly also the vector) interactions intervened in the decays and that in the muon decay the order of the factors written in the Fermi interaction was different from that for the electron. This was, however, also inconsistent with the *V-A* theory. Taylor (1958) found that he could also explain the slowness of the electron decay but only at the expense of also forbidding the muon decay, the dominant decay mode.

These theoretical speculations became moot when Steinberger and his collaborators (Impeduglia et al. 1958) found evidence for the electron decay of the pion. Using a liquid hydrogen bubble chamber they found six *clear* (emphasis in the original) examples of the $\pi-e$ decay. They stopped pions in the hydrogen and looked for events in which the stopped pion emitted a minimum ionizing particle (presumably an electron), with no visible intermediate μ meson track. Figure 5.4 shows examples of both types of event, one without a visible muon track (*A*) and one with such a track (*B*). They also measured the momentum of the secondary decay particle. Although most of the events selected were examples of $\mu-e$ decay with a very short muon track, the electron decays could be separated from these using a momentum criterion. The max-

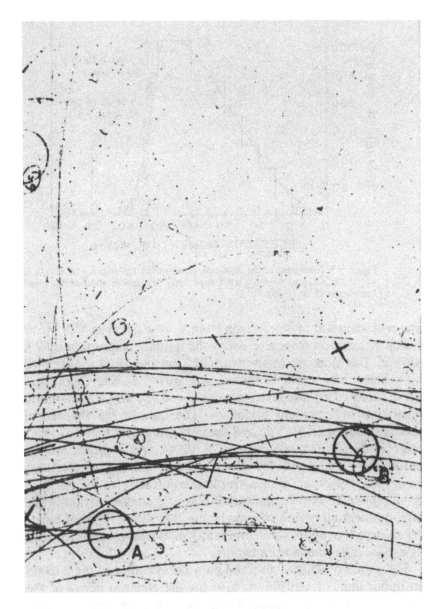

Figure 5.4. A bubble chamber photograph from Impeduglia et al. (1958). A π–e decay is shown at *A*. An example of π–μ–e decay is seen at *B*.

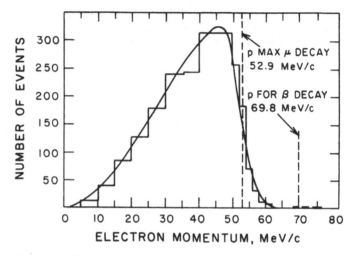

Figure 5.5. Histogram of the momenta of secondary particles for all events in which the incoming stopping track apparently decays into an electron. From Impeduglia et al. (1958).

imum momentum of an electron from $\pi-\mu-$e decay is 52.9 MeV/ c. For an electron from $\pi-$e decay the unique momentum is 69.8 MeV/c. The momentum spectrum of electrons from events with no visible muon track is shown in Figure 5.5. The theoretical curve assumes the Michel parameter $\rho = 0.75$, with a momentum resolution of 3 percent folded in. The data fit the theoretical spectrum below 65 MeV/c, and the curve indicates that no events are expected in the region near 70 MeV/c. As a check, the experimenters measured the momentum spectrum for electrons from the 2983 events with a visible muon track. No events were found above 62 MeV/c, indicating that $\mu-$e contamination in the high-momentum region was negligible. Decays in flight of either the pion or the muon were eliminated on kinetic grounds for each of the six high-momentum events.

Their total of 65,000 observed pion decays allowed them to set an upper limit of 1/10800 ± 40% for the electron decay of the pion. This was to be compared with the *V-A* prediction of 1/8000. They concluded, "The method does not yield a precise measurement of the branching ratio and cannot reasonably be extended to do so. However, the results here offer a very convincing proof of the existence of this decay mode, and show that the relative rate is close to that expected theoretically" (Impeduglia et al.

1958, p. 251). In an adjoining article, a CERN group (Fazzini et al. 1958) set a lower limit for the ratio of electron to muon decay of greater than 4×10^{-5}, "not in disagreement with that predicted by Ruderman and Finkelstein."

The V-A theory was also consistent with the then most recent results on radiative pion decay into an electron, $\pi \rightarrow e + \nu + \gamma$. Cassels et al. (1957) had measured this to be less than 1×10^{-5}. They noted, "the present limit has quite disastrous consequences for the simple universal Fermi interaction. This point has been thoroughly discussed by Treiman and Wyld" (Cassels et al. 1957, p. 734). The Universal Fermi Interaction discussed by Treiman and Wyld (1956) included the tensor interaction. That was the accepted view at the time. On the basis of this interaction they found that the branching ratio for radiative electron decay was greater than 0.025, inconsistent with the experimental value at that time of 5×10^{-5}, and even less consistent with that of Cassels et al. The consistency of the Cassels result with the V-A theory was noted by Sudarshan and Marshak (1958) in their short summary of the theory, published in 1958.

5.2c The asymmetry in the decay of polarized neutrons

In their original paper on the V-A theory, Sudarshan and Marshak (1957) had noted that the work of Telegdi and others, communicated privately, on the asymmetry in the β decay of polarized neutrons did not agree with their theoretical prediction. That result was published in late 1957 (Burgy et al. 1957). The angular distribution of decay electrons with respect to the neutron spin was $W(\theta) = 1 - (0.37 \pm 0.11) v/c \cos\theta$. The two-component neutrino theory (consistent with V-A) gave either -0.08 or -1.00, whereas parity conservation gave -0.21 or -0.86. The experimental results disagreed with both theories.

Subsequent analysis showed that the measured coefficient, given above, was too large in absolute value because of an interference from the correlation of the neutrino direction with the neutron spin. This error had been acknowledged in an invited paper given at the meeting of the American Physical Society held in January 1958 (Krohn 1957). A theoretical analysis by Jackson, Treiman, and Wyld (1957), performed before the experiment had been done, indicated that this coefficient might be large. Telegdi and his collaborators measured this correlation (Burgy et al. 1958). They found it to be $B = 0.88 \pm 0.15$. They also remeasured the correlation coefficient A, between the electron and the neutron spin, taking this value of B into account. They found $A = -0.09 \pm 0.03$, where the uncertainty was a lower limit. They concluded "that the interaction is predominantly V-A."

5.2d The polarization of the electron in muon decay

The fourth anomaly for the V-A theory was the observed polarization of electrons from muon decay. The theoretical argument was first given by Lee (1957) and goes as follows. "At the high-energy end of the beta spectrum [of muon decay] the neutrino and antineutrino are emitted together in the opposite direction from the beta particle. Because the neutrino and antineutrino spin in opposite directions their net spin is zero, therefore the β particle has the same spin as the muon. From the asymmetry in the beta distribution it is known that those states are strongly favored in which the β-particle momentum is antiparallel to the muon momentum in the pion rest frame. If the β particle is right-handed then the muon is left-handed, and vice versa" (Macq, Crowe, and Haddock 1958, p. 2062). If the neutrino in π^+ decay is left-handed then the μ^+ from such a decay is also left-handed. It then follows that the positron from this decay must be right-handed. Similarly, the electron from μ^- decay is left-handed.

At the 1957 Rochester conference, Lederman reported a tentative result inconsistent with the V-A theory. Using the circular polarization of bremsstrahlung from the positron from positive muon decay, he and his collaborators found that "the ratio of counts with their magnet one way to that with the magnet reversed was 1.06 ± 0.09. Conservation of leptons should give 0.95" (Lederman 1957, p. VII-31). Subsequent work by the same group was inconclusive (Garwin 1957).

The first of the new results was that of Culligan et al. (1957) on the polarization of positrons from μ^+ decay. Their method, which had already been used by Goldhaber, Grodzins, and Sunyar in an experiment on Eu^{152m} (discussed in detail in the next section) to establish the V-A theory, was as follows. The positrons from π–μ–e decay produced bremsstrahlung quanta in a lead plate. These quanta (γ rays) maintain the polarization of the incident positrons. The quanta then passed through an iron cylinder which could be magnetized either parallel or antiparallel to the motion of the quanta. The absorption of the γ rays is smaller in this energy range (a few tens of MeV) when the spins of the quanta are parallel to the electron spins in the material than when they are antiparallel. Thus one can determine the polarization of the rays, and thus of the incident positrons. The experiment measured the transmission of the quanta through iron with alternate magnetizations. They found a difference in counting rate for the two configurations of (4.54 ± 1.18) percent. The experimenters assumed that the beta decay interaction was scalar and tensor and concluded, "Thus the result of the experiment is contrary to the prediction of the two-component theory with lepton conservation. However, it appears that the scheme can be saved if the beta-decay interaction has been wrongly identified[10] (that is, if the interaction is vector and axial vector rather than scalar and tensor . . .)" (Culligan et al. 1957, p. 751).

A second paper by the same group (Culligan, Frank, and Holt 1959) included measurements on both positrons and electrons from both positive and negative muon decay. Their results for the integrated effect above 12 MeV were (4.7 ± 1.2) percent for μ^+, and (−5.6 ± 2.3) percent for μ^-. By this time the V-A theory was known to the experimenters and they noted, "This argument provides an important part of the evidence for the V and A types of coupling in β decay as opposed to the scalar S and tensor T types which were favored until recently (Culligan et al. 1957 [their previous experiment]; Sudarshan and Marshak 1958 [a reference to the published summary])" (Culligan, Frank, and Holt 1959, p. 171).

A similar experiment was done at approximately the same time by Crowe and his collaborators (Macq, Crowe, and Haddock 1958) and published somewhat earlier. They used the same method of absorption of bremsstrahlung in magnetized iron. They found an effect of (6.1 ± 0.9) percent for photons above 8 MeV

---

[10] Their footnote cited the new result of Hermannsfeldt et al. (1957) on the angular correlation in A^{35} that had indicated that V was preferred. No mention was made of the V-A theory.

from positrons, and (-4.9 ± 1.5) percent for electrons. This was to be compared to theoretical values between 3.5 and 6.3 percent for right-handed positrons and values between -6.3 and -3.5 percent for left-handed electrons. They concluded,

the results . . . indicate that β^+ is right-handed and β^- is left-handed. . . . The results are in agreement (or, more accurately, not in disagreement) with the assumption of (a) the two-component theory with left-handed neutrinos; (b) conservation of leptons; (c) universal β-decay theory with V and A interactions; (d) complete polarization of both β^+ and β^- (Macq, Crowe, and Haddock 1958, p. 2071).

5.2e The helicity of the neutrino: the Goldhaber-Grodzins-Sunyar experiment

By early 1959 all of the anomalies for the V-A theory had been resolved. Even before this, the theory had received strong support from an experiment by Goldhaber, Grodzins, and Sunyar (1958) that measured the helicity (handedness) of the neutrino. The physics of the experiment is as follows. Consider a spin-zero nucleus that decays by electron capture to an excited state of a daughter nucleus with spin one. This is a process in which the nucleus absorbs an orbital electron, transforms into a daughter nucleus with one less positive charge, and emits a neutrino. The resulting spin of the daughter nucleus must be antiparallel to the spin of the neutrino because of angular momentum conservation. If the neutrino is left-handed (axial vector interaction) then the nuclear spin must be antiparallel to the nuclear recoil direction. By conservation of momentum the recoil nucleus and the neutrino must come off in opposite directions. The γ rays emitted by the excited daughter nucleus in the direction of the nuclear recoil must carry this unit of angular momentum. They will therefore be left-circularly polarized in the case of a left-handed neutrino (A) and right-circularly polarized in the case of a right-handed neutrino (T). "Thus a measurement of the circular polarization of the γ rays which are resonant scattered by B [a target consisting of the same element as the daughter nucleus] yields directly the helicity of the neutrino, if one assumes only the well-established conservation laws of momentum and angular momentum" (Goldhaber, Grodzins, and Sunyar 1958, p. 1016). The resonant scattering guaranteed that the γ rays were emitted in the same direction as the recoil nucleus.

The experiment of Goldhaber, Grodzins, and Sunyar used a source of Eu^{152m} which decays to Sm^{152}. The γ rays from Sm^{152} passed through magnetized material, whose direction of magnetization could be alternated. As discussed above, the absorption of such γ rays depends on the relative orientation of the γ-ray spin and that of the electrons in the material. Thus, the helicity of the γ rays could be determined. The γ rays were then scattered by an Sm_2O_3 target, which would ideally scatter only those γ rays emitted in the recoil direction. They found a difference between the two magnetizations of 0.017 ± 0.003, after nonresonant background was subtracted. This was to be compared with an expected value of ± 0.025 for 100% circularly polarized γ rays, with the minus sign corresponding to positive helicity (T) and the plus sign to negative helicity (A). They concluded, "our result indicates that the Gamow-Teller interaction is axial vector (A) for positron emitters" (Goldhaber, Grodzins, and Sunyar 1958, p. 1017).

By early 1959 all of the experimental evidence from weak interactions – nuclear beta decay, muon decay, pion decay, and electron capture – was in agreement with the V-A theory.[11] As Marshak remarked later, "And so it came to pass – only three years after parity violation in weak interactions was hypothesized – that the pieces fell into place and that we not only had confirmation of the UFI [Universal Fermi Interaction] concept but we also knew the basic (V-A) structure of the charged currents in the weak interactions for both baryons and leptons" (Sudarshan and Marshak 1985, p. 14).

5.3 DISCUSSION

What can one conclude from the history of this episode recounted in these five chapters? It certainly illustrates the fallibility and corrigibility of experimental results. During the 1930s the accepted spectral results changed considerably as scientists found that the nature of the radioactive source affected the observed spectrum. In the early experiments scattering and energy losses

---

[11] A problem remained because the coupling constants G_A and G_V were not, in fact, equal. This was later explained as being due to renormalization effects. It did not seem to affect the acceptance of V-A as a universal theory.

in the source resulted in too many low-energy electrons and in too low an average energy. Experimental difficulties also affected the high-energy end of the spectra. The later experiments, done with thin sources, resulted in very different spectra.

Even after the problem had apparently been solved, difficulties appeared in the beta-decay spectra observed in the late 1940s and early 1950s. The observed spectra again seemed to differ from the Fermi prediction. Once again, source thickness was the culprit. Even sources that appeared to be uniform to the naked eye showed thickness variations of a factor of 100, which affected the experimental results. More careful source preparation helped to solve the problem.

One can also point to the original experimental anomalies for the V-A theory, which were subsequently resolved by the repetition of the experiments and more careful analysis.

Difficulties can also attend the comparison of experimental results with theory. It is very rare that raw data can be directly compared with theoretical prediction. Theoretical analysis and calculation is almost always needed. Thus, in the analysis of the RaE spectrum, which led to the inclusion of the pseudoscalar form (P) in the beta-decay interaction, Petschek and Marshak used an incorrect finite radius correction. A more accurate correction changed their conclusion. Similarly, more rigorous Coulomb corrections to the beta-decay spectra helped to resolve the difficulties in the 1950s. In the decay of polarized neutrons the experimenters overlooked a correlation of the neutrino direction with the neutron spin. Only when that correlation was measured and taken into account could a correct result be obtained.

A second point involves the difficulty that sometimes attends the articulation of a theory so that it can be tested. As we have seen, the Fermi theory for allowed spectra was tested by comparing its predictions to experimentally observed forbidden spectra. Only after the theoretical spectra for forbidden transitions were calculated by Konopinski and Uhlenbeck could a valid test be made and the discrepancy resolved. Ironically, this calculation supported arguments against Konopinski and Uhlenbeck's own modification of Fermi's theory.

As experimental results change, the support, or lack of it, which they provide for different theoretical explanations also changes – as one would expect. Confirmation and refutation are also, nec-

essarily, fallible and corrigible because they are based on experimental outcomes. For example, the early spectra refuted Fermi's
theory and confirmed the Konopinski-Uhlenbeck modification.
After those results were found to be in error and corrected, that
decision was reversed.

Does this fallibility and corrigibility argue against a legitimate
role for experiment in theory choice? I think not. It shows only
that the choices are themselves fallible and corrigible, and not
that they were not reasonable. (This issue will be discussed in
detail in the next chapter). Instant and permanent rational choice
does not happen in science. Experimental difficulties, problems
with theoretical analysis of the data, and incorrect theoretical
comparisons can, and do, occur. What this history shows is that
such errors are detected and corrected, without waiting for the
adherents of the older view to die off. The decisions are made
on the basis of the best experimental evidence available. Thus,
it was only six years between Konopinski and Langer's statement
that *STP* was the form of the weak interaction and the acceptance
of the *V-A* theory which contradicted it. It was only eight years
between the formulation of the Konopinski-Uhlenbeck theory,
its apparent confirmation and acceptance, and its refutation on
the basis of the experimental evidence, as noted by Konopinski
himself. In the case of parity nonconservation, it seems fair to
say that as soon as physicists saw the experimental evidence, they
were willing to give up a strongly held belief in an accepted conservation law.[12] Parity nonconservation is still accepted.

Another question concerning the use of experimental results
in theory choice is whether or not the theoretical beliefs or presuppositions of the experimenters have an effect on what they
observe or report. This comes to mind particularly with regard
to the four experimental anomalies for the *V-A* theory and their
subsequent removal. When *STP* was the accepted theory the
anomalous results were reported, which were consistent with the
STP interaction. After the suggestion of *V-A* and its initial support, the anomalous results were changed. This raises at least the
possibility that the theoretical commitments of the experimenters
affected their results. I believe that the history recounted above
demonstrates that this was not the case; in this episode, the the-

---

[12] See Franklin (1986, ch. 1).

oretical presuppositions had no effect on the experimental results.[13]

Let us consider each of the anomalous cases separately. In the case of the angular correlation in He[6] it should be remembered that it was the Rustad-Ruby result that strongly supported tensor (T) as the form of the Gamow-Teller interaction, and not vice versa. There was also a known place to look for a possible error in the experimental results. Rustad and Ruby had assumed that all of the observed decays came from gas contained in the target volume. They, themselves, recognized that this assumption was questionable, although their calculation of their results used it.[14] If this assumption, that all the decays came from the source region, was not correct then the results would have to be modified. Wu and Schwarzschild investigated the question and showed that the assumption was not, in fact, correct and that there was a considerable amount of gas in other parts of the apparatus. Thus, the Rustad-Ruby results were questionable. The repetition of the experiment by Allen and his collaborators (Hermannsfeldt et al. 1959) changed the conclusion.

Perhaps the second most important anomaly was the failure to observe the electron decay of the pion. This was inconsistent not only with the V-A theory but with all other proposed theoretical explanations at the time. It seems difficult to imagine how the theoretical beliefs of the experimenters could have influenced the results. Beliefs, no matter how fervent, cannot put events into a bubble chamber. One may wonder why the decays were not seen previously. Here, too, it is difficult to see any effect of presuppositions. The failure to observe the decays was a problem for all theories. These were, after all, rather rare events and might easily have been missed.

In the case of the decay asymmetry from polarized neutrons the experimenters had initially overlooked a correlation between the neutrino direction and the neutron spin that affected their result. This correlation had been suggested prior to both the experiment and to the suggestion of the V-A theory. It was not invented for the purpose of explaining the anomaly. When the

---

[13] In the next chapter I will discuss other cases which are more problematical.
[14] Dr. S. Ruby (private communication).

effect was measured and included in the calculations the anomaly disappeared. The anomaly in the decay of the muon is even less problematical. The rather small anomalous effect appeared only in a preliminary result. Later results by the same group were inconclusive and were superseded by more accurate and precise experimental results. One should also note that these later results were initially regarded as anomalous for the then accepted *STP* model.

Despite the initial worry, there seems to be no evidence that theoretical presuppositions or beliefs played any role in the removal of the experimental anomalies for the *V-A* theory. The initial support for the theory certainly caused the reexamination and repetition of the experiments, but this is as it should be. As we have seen, experimental results are fallible.

We have also seen that both the experimenters themselves, and the physics community in general, seem quite prepared to accept results in disagreement with an accepted theory. Results in disagreement with the original Fermi theory, with its successor, the Konopinski-Uhlenbeck theory, and once again with the Fermi theory, were observed in the 1930s and 1940s. The results that demonstrated the nonconservation of parity seem to have been accepted immediately, even though parity conservation was a strongly believed conservation law. The experimental results of Hermannsfeldt et al. (1957) on the angular correlation in A^{35}, which argued for the presence of a vector interaction, disagreed not only with the then accepted theory, but also with other accepted experimental results.

Although this history has concentrated on the role of experiment in testing, confirming, or refuting theories, we have also seen another role that experiment plays. This is in the acquisition of data that are believed to be of use for future theoretical explanations. Both Sargent's work on beta decay and the large amount of effort expended on measuring the muon decay spectrum were done in the absence of any theory of the phenomena. There are also other examples of what Hacking has referred to as experiment having a life of its own, such as the investigation of K mesons, leading to the θ–τ puzzle, and the initial angular correlation experiments.

In the next chapter I will present a philosophy of experiment.

I will include a discussion of the issues raised concerning the role of fallible experimental results in theory choice. I will also examine the question of how we come to believe reasonably in experimental results, and thus justify their role in theory choice.

II

TOWARD A PHILOSOPHY
OF EXPERIMENT

"Empiricism is more fun than speculation."

S. Washburn (1986)

6

Experimental results

A skeptical reader, after reading the history presented in the first five chapters, might question whether or not science is the reasonable enterprise based on valid experimental evidence that I believe it is. They might say that the demonstrated fallibility of experimental results, of theoretical calculation, and of the comparison between experiment and theory casts serious doubt on that assertion. I agree that the fallibility is worrisome, but I also believe that the corrigibility shown gives us hope that science is indeed a reasonable enterprise. We not only learn from our mistakes, but we are able to correct them. In this chapter I will argue that we have good reasons for belief in the validity of experimental results.[1] I will discuss later how we can use the results in the construction of a reasonable, dare one say, rational, science.[2]

6.1 THE BAYESIAN APPROACH TO THE PHILOSOPHY OF SCIENCE

I will adopt a Bayesian approach to the philosophy of science. Bayesianism is based on the idea that we have degrees of belief in statements or hypotheses, and that these degrees of belief obey the probability calculus. [For an excellent introduction to the Bayesian view see Howson and Urbach (1989).] There has been considerable discussion of what kind of probabilities these are. I

---

[1] As discussed later, the validity of an experimental result does not guarantee that the result is correct.
[2] As mentioned earlier, I will use the term reasonable, rather than rational, in this discussion because I do not have a complete theory of rationality. I do believe, however, that when you observe evidence entailed by a hypothesis it is rational to strengthen your belief in that hypothesis. I will argue later that the Bayesian view of science, which incorporates this, is the correct one.

believe that they are subjective probabilities reflecting the judgments of scientists. I am not suggesting that scientists are conscious Bayesians, but as will be discussed later, the Bayesian approach can explicate both various episodes in the history of science and the strategies used to validate experimental results. Scientists seem to behave as if they were Bayesians, even if they don't calculate, or even estimate, the probabilities explicitly.[3]

There is some anecdotal evidence to support the view that scientists do make such probability estimates. When the experiments that ultimately demonstrated that parity conservation, or space reflection symmetry, was violated were being planned, several scientists gave estimates of the likelihood of positive results. Richard Feynman told the following story. "Norm Ramsey asked me if I thought he should do an experiment looking for parity law violation and I replied, 'The best way to explain it is, I'll bet you fifty to one you don't find anything' " (Feynman 1985, p. 249). Frauenfelder and Henley report, "Feynman paid" (1975, p. 389). Wolfgang Pauli wrote to Victor Weisskopf, "I do not believe that the Lord is a weak left-hander, and I am ready to bet a large sum that the experiments will give symmetric results" (quoted in Bernstein 1967, p. 59). Fortunately for both Pauli and Felix Bloch, who offered to bet his hat (T. D. Lee, private communication), no one took them up on these bets.

The underlying mathematical statement of the Bayesian approach is known as Bayes's Theorem. It states that

$$P(h \wedge e) = P(h|e) P(e) = P(e|h) P(h),$$

where $P(h|e)$ is the conditional probability of h given e, and $P(e)$ and $P(h)$ are the prior probabilities of e and h, respectively. These probabilities are all assumed to be relative to some background knowledge.

When h entails e, $h \vdash e$, $P(e|h) = 1$. This gives the intuitively appealing result that the observation of evidence entailed by a hypothesis strengthens our belief in that hypothesis. In this case, $P(h|e) = P(h)/P(e)$. Because $P(e)$ is always less than 1, $P(h|e) >$

---

[3] Some psychological studies have shown that scientists tend to represent likelihoods in categorical terms such as low, moderate, or very high or in terms of ordinal relations such as greater than, less than, or equal to, rather than in numerical terms (Kuipers, Moskowitz, and Kassirer 1988).

$P(h)$. Using the usual Bayesian support function, Support = $P(h|e) - P(h)$, this means that h is supported by e.[4]

A question might be raised as to what I mean by these prior probabilities, particularly when this analysis is applied to historical episodes. These priors are my own estimate of what a typical scientist, at the time, would have given as an estimate of the probability. They are not the actual or estimated judgment of any particular scientist. One objection to this approach might be that if different scientists have different judgments as to the prior probabilities of a hypothesis, as they do, then even if they conditionalize on the evidence they will not agree on the posterior probabilities, $P(h|e)$.[5] Therefore, given these different priors a consensus will not develop. This would certainly be at odds with the actual history of science, where we do see scientists agreeing on what they view as the best theory or hypothesis. The posterior probabilities may differ, but I believe that in the practice of science the estimates of the priors do not differ by enough to prevent the scientific community from agreeing on the best confirmed or supported hypothesis, given a reasonable amount of evidence.[6]

A case in point is the discovery of CP violation. In this case the combined symmetry of charge conjugation, or particle–antiparticle symmetry, and parity, or space reflection symmetry, was shown to be violated. (See Franklin 1986, ch. 3, for details.) In 1964, an experiment by Christenson and collaborators (1964) demonstrated the existence of $K^{\circ}_2 \rightarrow \pi^+\pi^-$ decay. The most obvious interpretation of this result was that CP symmetry was violated, or apparently violated. This was the view expressed in three out of four theoretical papers on the subject published by the end of 1965. Alternative explanations were also offered. These included the following: The decay particles were not K°_2 mesons,

---

[4] A Bayesian is not committed to accepting the hypothesis with the highest posterior probability, $P(h|e)$. After all, a hypothesis that merely restates the evidence has a probability of 1. Acceptance depends on the purpose for which the hypothesis or theory will be used. In general, it will involve the scope of the theory, its perceived fruitfulness, etc.

[5] I regard this as a strength rather than a weakness of the Bayesian approach because it can accommodate the different judgments of scientists. Bayesianism tells us how we should change our beliefs in the light of evidence, not what our prior degrees of belief should be.

[6] I am referring here to nonstatistical hypotheses. For statistical hypotheses, Savage (1982) has shown that in the limit of infinite evidence the conditionalized probabilities will approach the actual frequencies.

the decay particles were not pions, another particle was emitted in the decay, the exponential decay law was violated, the principle of superposition in quantum mechanics was violated, there existed a shadow universe, and others. Within three years all of these alternative explanations had been tested and found wanting, leaving only *CP* violation as the explanation. Even before this occurred, however, the physics community had, by the end of 1965, reached a consensus that *CP* was violated, even though all the alternatives had not been tested. The untested explanations remaining were regarded as so unlikely, relative to what was accepted by the physics community (background knowledge), that they did not need to be tested before a consensus was reached. I will return to this point later, when I discuss theory choice.

The Bayesian approach to science has had success in explicating various aspects of science. Thus, Good (1967) and Maher (1988) have shown, using expected utilities and Bayesian decision theory, that it is reasonable for scientists to gather data.[7] The intuitively appealing result that two "different" experiments provide more confirmation of a hypothesis than repetitions of the "same" experiment has also been demonstrated using Bayesian methods (Urbach 1981; Horwich 1982; Franklin and Howson 1984). On the historical side there have been Bayesian accounts of Newton's derivation of the inverse square part of his law of universal gravitation (Franklin and Howson 1985), and the resolution of anomalies for Dirac theory in the double scattering of electrons during the 1930s (Franklin 1986, ch. 4). On the philosophical side, successful solutions have been proposed for two of the traditional problems of Bayesianism: (1) that of known evidence and (2) the "tacking" paradox or the localization of support.

Some critics have claimed that a Bayesian cannot account for the fact that evidence known before a hypothesis was suggested can be used to support the hypothesis. Because this seems to violate both common sense and the history of science, it is a telling argument against the Bayesian view. After all, Kepler's laws were known well before Newton formulated his inverse square law of gravitation, and were, and are, regarded as extremely strong support for the law. The critics note that if the evidence is known

---

[7] This may be an obvious conclusion, but it is useful to provide an argument in its favor.

then a Bayesian must regard $P(e) = 1$, and thus that $P(h|e) = P(h)$, and the evidence cannot support the hypothesis.[8] Howson (1985, 1987) has suggested that one should eliminate e from our background knowledge, thus making $P(e) \neq 1$, and say that our new belief in h, $P(h|e)$ is counterfactually what our belief in h would be were we to come to know e.[9]

The "tacking" paradox has also been a problem for the Bayesian view. Quantum mechanics entails the Balmer series of hydrogen. Thus, observation of the Balmer series confirms quantum mechanics. If, however, one adds to, or tacks onto, quantum mechanics an additional statement, s, such as "The moon is made of green cheese," then the conjunction of quantum mechanics and s also entails the Balmer series and observation of the series will confirm the conjunction. This seems somewhat counterintuitive. A Bayesian solution has also been proposed for this problem. Colin Howson and I (Howson and Franklin 1985) have shown that the support for the conjunction is reduced by a factor that depends on the excess content of that conjunction. In the case of tacking on a known falsehood such as "The moon is made of green cheese" the support for the conjunction is, in fact, zero.[10]

6.2 IT PROBABLY IS A VALID EXPERIMENTAL RESULT: A BAYESIAN APPROACH TO THE EPISTEMOLOGY OF EXPERIMENT[11]

I have previously argued (Franklin 1986, chs. 6, 7) that there is an epistemology of experiment, a set of strategies that pro-

---

[8] The support function I am using is $P(h|e) - P(h)$. This measures the change in our belief due to the acquisition of the evidence.

[9] Niiniluoto (1983) has also given a somewhat different Bayesian solution to this problem.

[10] The general result is that if $t \vdash h \vdash e$ then the support of t by e, $S(t,e) = P(t|e) - P(t) = P(t|h) S(h,e) = P(t|h) [P(h|e)-P(h)]$. $P(t|h)$ is a decreasing function of the excess content of t with respect to h. Some of the difficulty seems to be that support of a conjunction seems to imply support of either conjunct. Thus, one might feel that observation of the Balmer series supports the statement "The moon is made of green cheese." This is certainly not the case. As Miller and Popper (1983) have shown, evidence that supports a statement does not necessarily support all the logical consequences of that statement.

[11] Much of the work in this section was done in collaboration with Colin Howson. I am grateful to him for allowing me to include it here.

vides reasonable belief in the validity of an experimental result. These strategies distinguish between a valid observation or measurement[12] and an artifact created by the experimental apparatus. These strategies, discussed in detail subsequently, include the following:

1. Experimental checks and calibration, in which the apparatus reproduces known phenomena.
2. Reproducing artifacts that are known in advance to be present.
3. Intervention, in which the experimenter manipulates the object under observation.
4. Independent confirmation using different experiments.
5. Elimination of plausible sources of error and alternative explanations of the result.
6. Using the results themselves to argue for their validity.
7. Using an independently well-corroborated theory of the phenomena to explain the results.
8. Using an apparatus based on a well-corroborated theory.
9. Using statistical arguments.

In this section I will show that these strategies can be incorporated, quite naturally and plausibly, within a Bayesian approach to the philosophy of science. This satisfies an important criterion for a good philosophy of science, that it explicate in a reasonable way the accepted good practice of science. As the examples will illustrate, these strategies are used by practicing scientists.

I now present Bayesian accounts of the strategies used to validate experimental results. The first three strategies are simple instances of the fact that observation of evidence entailed by a hypothesis strengthens our belief in the hypothesis. Here the hypothesis is "The experimental apparatus is working properly."

The most important and widely used strategy is that of experimental checks and calibration. The experimenter checks that the apparatus can reproduce known results. Thus, for example, if one wished to establish that the newly observed spectrum of a substance is valid one might check that the spectrometer can correctly reproduce the known Balmer series of hydrogen. If the Balmer series is observed correctly then our belief would be strengthened

---

[12] I will be concerned here, primarily, with observations or results that are interpreted within an existing theory and not with pointer readings, etc.

that the apparatus is working properly and in other observations or measurements made with that apparatus.

Another method of checking that an apparatus is working properly is to see whether or not it reproduces artifacts known in advance to be present. An illustration of this concerns the observation of the infrared spectra of organic compounds (Randall et al. 1949). In many cases a pure sample of the compound could not be prepared and the compound had to be observed in an oil paste or in solution. One then expects to observe the spectrum of the oil or the solvent superimposed on the spectrum of the compound. The observation of these spectra thus provided a check on the proper operation of the apparatus and may also provide a calibration (Figure 6.1).[13]

Similarly, this approach accounts for Hacking's (1983) strategy of intervention. Hacking's example is the microscope, in which the experimenter stains the material, injects fluid, or in other ways manipulates the object under observation. The observer predicts what will be observed after the intervention, that is, the object will change color after it is stained. When the predicted behavior is seen then we strengthen our belief in the proper operation of the microscope and in other observations made with it. A particularly dramatic instance of this strategy is that the observed effect will disappear under certain circumstances. Thus, in one of the experiments demonstrating the nonconservation of parity, Wu et al. (1957) predicted that the observed asymmetry in the β decay of oriented nuclei would disappear as the sample warmed up and the orientation of the nuclei was destroyed (Figure 6.2). The disappearance of the asymmetry gave increased confidence in its validity.

Of course, if such a check fails, then our belief in the proper operation of the apparatus and in observations made with it is weakened, if not entirely destroyed. An illustration of this occurred in an experiment that measured the spin and magnetic moment of He^6 (Commins and Kusch 1958). In that experiment

---

[13] An interesting corollary to this strategy is suggested by the K^+_{e2} experiment discussed later in this chapter. In that experiment an unexpected artifact appeared during the test of the beam line. The reason for the artifact was found and the explanation increased the experimenter's confidence in their apparatus. If you can explain an artifact then you strengthen your belief in the proper operation of your experimental apparatus.

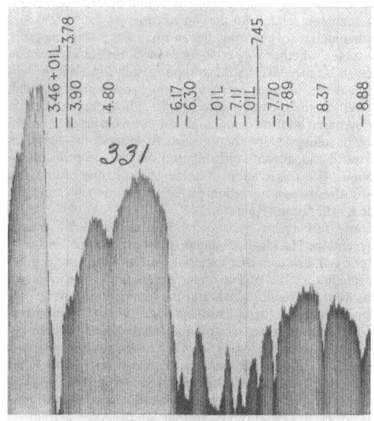

PLATE 58.   Assignments: 6.17 μ   δNH₃⁺
                       6.30 μ   Carboxylate ion
                 Preparation: Oil paste

Figure 6.1. Infrared spectrum of an organic molecule prepared in an oil paste.
The oil spectrum is clearly indicated. From Randall et al. (1949).

a beam of He^6 atoms passed through a long, inhomogeneous magnetic field. If the spin of He^6 were 1 then the ratio of counts observed with the magnet off to those with the magnet on should have been three to one. This was observed in the first experimental run. An experimental check was performed by destroying the collimation of the beam so that the atoms did not pass through the magnetic field. Under those circumstances the counting rates with the magnet off and with the magnet on should have been

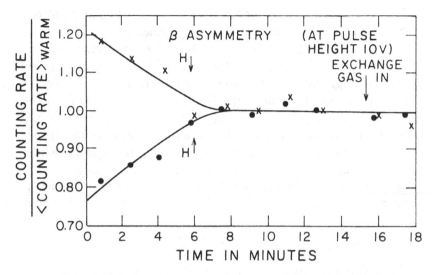

Figure 6.2. The decay electron counting rate as a function of time for the decay of polarized Co⁶⁰ nuclei. Note that the asymmetry between the two polarizations disappears as the sample warms up. From Wu et al. (1957).

equal, if the apparatus had been working properly. The factor of three difference persisted when the check was done, indicating that something was wrong with the operation of the apparatus. In fact, the magnetic field had reduced the efficiency of the detector by an unfortunate factor of three. Better magnetic shielding of the detector corrected the problem, the experimental check was successfully performed, and the spin of He⁶ was found to be zero.[14]

Another method of establishing the validity of an experimental result is by independent confirmation, the use of two "different" experiments. Hacking's illustration is of dense bodies in cells, which were observed using different microscopes, that is, optical, phase contrast, interference, and so forth. The fact that the same phenomenon was observed under very different conditions makes it extremely unlikely that it is an artifact created by the experimental apparatus.

This is a particular instance of the general result that a hypothesis receives more confirmation from two "different" experiments than from repetitions of the "same" experiment (Franklin and Howson 1984). Thus, if we wish to know the correct time, it

[14] I was an undergraduate assistant on the experiment.

is better if we compare watches than if either of us looks at our own watch twice. Let us consider a hypothesis h, which for the case of experimental results is of the form "X has been observed" or "The value of X is a." Let E and E' be two experimental apparatuses giving rise to results e and e', respectively. We say that e and e' are the results of "different" experiments if, at the $k + 1$ trial, $P(e'_{k + 1}|e_1 \wedge e_2 \wedge \cdots \wedge e_k) < P(e_{k + 1}|e_1 \wedge e_2 \wedge \cdots \wedge e_k)$, and vice versa. The conditions under which one makes such probability assignments involve both the theory of the apparatus and the theoretical context.[15] It then follows quite easily that h receives more confirmation from e and e' than from repetitions of either. One can show that

$$\frac{P(h|e_1 \wedge e_2 \wedge \cdots \wedge e_k \wedge e'_{k+1})}{P(h|e_1 \wedge e_2 \wedge \cdots \wedge e_k \wedge e_{k+1})} = \frac{P(e_{k+1}|e_1 \wedge e_2 \wedge \cdots \wedge e_k)}{P(e'_{k+1}|e_1 \wedge e_2 \wedge \cdots \wedge e_k)}.$$

If the experiments are "different," then $P(e_{k + 1}|e_1 \wedge e_2 \wedge \cdots \wedge e_k)$ $> P(e'_{k + 1}|e_1 \wedge e_2 \wedge \cdots \wedge e_k)$ and $P(h|e_1 \wedge e_2 \wedge \cdots \wedge e_k \wedge e'_{k + 1}) > P(h|e_1 \wedge e_2 \wedge \cdots \wedge e_k \wedge e_{k + 1})$ as required.

A variant of the use of independent experimental apparatuses as an epistemological strategy is the strategy of indirect validation. Suppose we have an observation that can be made using only one kind of apparatus. Let us also suppose that the apparatus can produce other similar observations which can be corroborated by other techniques. Agreement between these different techniques gives confidence not only in these similar observations but also in the ability of the first apparatus to produce valid observations. This, then, provides an argument in support of the observation made only with that apparatus. An example of this is the observation of the microtrabecular lattice in cells using electron microscopy. It is argued that other objects of similar size (e.g., microtubules) have been seen by both electron microscopy and with an ordinary light microscope. This supports the idea that electron microscopy can detect objects of this size, and also helps to validate the observation of the lattice. Here the similarity of the observations, in this case in size, is of importance. The ability of the electron microscope to detect objects of very different size, particularly of larger size, would not be of much assistance in

---

[15] The theory-ladenness of observation, to be discussed later, is a virtue here, rather than a problem.

arguing for the lattice, because the well-corroborated theory of the apparatus indicates that size is an important parameter.

We can also establish the validity of an experimental result using what we might call the "Sherlock Holmes" strategy. As Holmes remarked to Watson, "How often have I said to you that when you have eliminated the impossible, whatever remains, *however improbable,* must be the truth" (Conan Doyle 1967, p. 638). The Bayesian tinge to Holmes's thought is clear. If we can eliminate all plausible sources of error and alternative explanations then we are left with a valid experimental result. Thus, when electronic signals were observed by Voyagers 1 and 2 the following possible causes were eliminated: defects in the telemetry, interaction of the spacecraft with the environment of Saturn, lightning, dust, and so on (Warwick 1981, 1982). What was left was a valid observation of electric discharges in the rings of Saturn.

A Bayesian approach shows quite clearly why this strategy works. Let H be the hypothesis of electric discharges in the rings of Saturn, e the observed signals, and h_i the plausible sources of error and alternative explanations of e.

$$P(H|e) = P(e|H)P(H)/P(e)$$
$$= P(e|H)P(H)/[P(e|H)P(H) + \sum_i P(e|h_i)P(h_i)].$$

Now let us suppose that we acquire additional information e_1 that eliminates h_j, that is, $h_j \vdash \neg e_1$ so $P(e_1|h_j) = 0$. In the example of Saturn this might be the check of the telemetry system by inserting a known signal into the apparatus. When that signal is received correctly on earth we can eliminate the hypothesis that the observation was due to a fault in the telemetry. Now

$$P(H|e \wedge e_1) = P(e \wedge e_1|H)P(H)/[P(e \wedge e_1|H)P(H) + \sum_{\substack{i \\ i \neq j}} P(e \wedge e_1|h_i)P(h_i)]$$

$$> P(H|e).[16]$$

---

[16] I am assuming here that e_1 is independent of H, of the remaining h_i, and of e. Thus, each term in the expression for $P(H|e \wedge e_1)$ will contain $P(e_1)$, and $P(H|e \wedge e_1)$ will be greater than $P(H|e)$ because the term containing h_j is removed from the sum. This assumption of independence is quite reasonable. For example, the observation of the signal used to check on the proper operation of the telemetry is independent of whether or not there are electric discharges in the rings of Saturn, and independent of the other alternative explanations.

We can continue this procedure, acquiring additional pieces of information $e_2, e_3, \ldots, e_N$, which eliminate all the competing h_i with any significant prior probability, the plausible sources of error and alternative explanations. We are left with

$$P(H|e \wedge e_1 \wedge e_2 \wedge \cdots \wedge e_N) = \frac{P(e \wedge e_1 \wedge e_2 \wedge \cdots \wedge e_N | H)P(H)}{P(e \wedge e_1 \wedge e_2 \wedge \cdots \wedge e_N | H)P(H) + \beta},$$

where β is the very small remainder of the sum.

Thus, $P(H|e \wedge e_1 \wedge e_2 \wedge \cdots \wedge e_N) \approx 1$. By a process of elimination, H has received considerable support.

A very similar Bayesian analysis can also explicate the fact that sometimes the results themselves may argue for their own validity. Consider Millikan's observations that the charges on oil drops were always integral multiples of a certain unit of charge. Although we might very well believe that Millikan's apparatus might give incorrect values of the charge on the drops, there was no remotely plausible malfunction of the apparatus that would give rise only to integral multiples of a single unit. Similarly, though one might believe, as Cremonini and other seventeenth-century Aristotelians are reported to have believed, that Galileo's telescope could create specks of light, it is extremely unlikely that it could create specks of light that behave as a planetary system with consistent motions and eclipses. It would be even more astounding if such artifacts obeyed Kepler's Third Law, $R^3/T^2 = $ constant. Thus, one supported the validity of the observation of the moons of Jupiter.[17] What one is really saying in these cases is that $P(e|h_i) << 1$, where h_i are plausible sources of error, so that, as above, $\sum_i P(e|h_i)P(h_i)$ is very small.

The case in which the validity of an experimental result can be established, at least in part, by the fact that the observation can be explained by an existing well-corroborated theory of the phenomena can also be explicated in a similar manner. This, of course, depends on the detailed nature of the theoretical prediction. If the prediction is rather vague this strategy is unlikely to be convincing. If the prediction is complex and specific then it does work. A case in point is the recent discovery of the W particle (Arnison 1983; Banner 1983), the intermediate vector boson predicted by the well-confirmed Weinberg-Salam unified theory of electromagnetic and weak interactions. The theory

---

[17] This is not a historical argument, but rather one that could have been given.

predicted not only the existence of such a particle but also predicted that it would decay into electrons with high transverse momentum relative to the beam producing the W, in events with a large amount of missing energy. The mass of the W and the transverse momentum distribution of the W bosons produced were also predicted. Although one might believe that background processes could give rise to such electrons (even that was considered very unlikely) it becomes very implausible that such processes or an equipment malfunction would result in the predicted decay mode, the predicted mass, and the predicted angular distribution. Thus, we are left with a valid observation of the W. This was not, of course, the only strategy used in the experiment. Parts of the apparatus were checked independently by seeing if they reproduced known results or predicted observations. Background processes that might have simulated W decay were also eliminated.

If the apparatus itself is based on a well-corroborated theory then we also have good reason to believe in its results. The support for the underlying theory increases our belief that the apparatus will work properly. This is the case for both the radiotelescope and the electron microscope. Of course, here too, other strategies may be used in conjunction with the theory of the apparatus to help validate an observation. In the case of the report of the electron microscope observation of a microtrabecular lattice in cells it was not the proper operation of the electron microscope that was doubted. Questions were raised whether or not the sample preparation had produced an artifact. Strategies such as independent confirmation, various experimental checks, and intervention were also used (Franklin 1986, ch. 6).

Bayesian analysis also helps us to understand how statistics were used to discover new elementary particles.[18] During the 1960s and 1970s high-energy physicists searched for such short-lived particles by examining graphs of the number of observed events plotted against the invariant mass of the decay particles (Figure 6.3). The existence of such a particle was signaled by a bump in the smooth distribution expected from other processes if no such particles were present. The usual informal criterion for the presence of a bump in a given bin was that it be three standard deviations (s.d.) above the background. The probability of a 3-s.d. effect in any given bin, if the events were distributed according to chance, was 0.27%. Thus, in any single experiment, the probability of observing

---

[18] I am grateful to Peter Galison for calling my attention to this strategy.

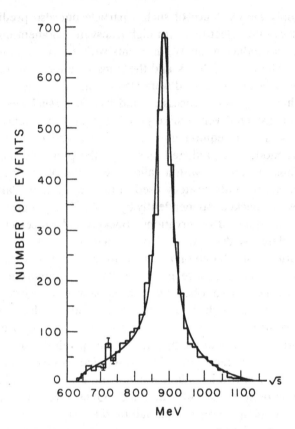

Figure 6.3. Distribution of masses for the K^0 system obtained by Wojcicki et al. (1963). Note that almost all the events are due to the presence of a new particle, the K^*, and that there is almost no background.

such an effect was quite small, if no particle was present. It was pointed out, however, by Arthur Rosenfeld,[19] that given the large number of such graphs plotted each year by physicists, the probability of observing a 3-s.d. effect was quite high even if the events were randomly distributed. The probability of a 3-s.d. effect in 1000 bins is 93%.[20] The informal

[19] This story may be apocryphal, but it did have wide circulation among high-energy physicists at the time.

[20] Another example may help to clarify this point. Suppose we have ten people each tossing a fair coin one hundred times. We would not expect to find one of them observing seventy-one heads (a 3-s.d. effect) very often. The probability is 2.7%. If we had 1000 people tossing coins then we would expect at least one of them to observe seventy-one heads. The probability of this is 93%.

criterion was then changed to requiring a 4-s.d. effect, which had a probability of 0.0064%, in any given bin, and which has a probability of 6% in 1000 bins.

We can understand this as follows. Let *h* be the hypothesis of a particle in the mass distribution, let *c* be the hypothesis that the distribution is due to chance, and let *e* be the observation of a 3-s.d. effect at m_o.

$$P(h|e) = P(e|h)P(h)/P(e).$$
$$P(c|e) = P(e|c)P(c)/P(e).$$
$$P(h|e)/P(c|e) = P(e|h)P(h)/P(e|c)P(c).$$

The judgment of the physics community was that although $P(e|h)P(h)$ was rather small (after all there are only approximately one hundred such particles known) it was still larger than $P(e|c)P(c)$ for any single experiment because $P(e|c)$ was so small for a 3-s.d. effect. Thus, the observation of such an effect was evidence for the existence of a new particle for any one experiment. But, in fact, the data implicitly refer to a sample space containing a much larger number of experiments. Now the probability of observing a 3-s.d. outcome conditional on *c* is quite high. So, quite correctly, Rosenfeld argued that one should not consider only a single experiment and its graphs, but all such experiments done in a year. This made the probability of observing a 3-s.d. effect considerably larger. Changing the criterion to 4 s.d. lowered that probability considerably, as we saw earlier.

The case becomes even clearer when a theory *T* predicts the existence of a particle at m_o. Then $P(e|T) = 1$ and all that is required for $P(T|e) > P(c|e)$ is that $P(T)/P(c) > P(e|c)$, which, for a 4-s.d. effect, is obviously not a very stringent condition because $P(e|c) = 0.000064$.

I am not suggesting that this set of strategies for validating experimental results is either exclusive or exhaustive.[21] Nor am I claiming that any single strategy or subset of strategies is either necessary or sufficient for such validation. Even though there may be, and no doubt are, other strategies for validating an experimental result and that whether or not any single strategy or set of strategies will suffice to establish validity depends on the particular experiment, I believe that these are reasonable strategies,

---

[21] Since I began working on this question I have added several strategies to my original set.

and do serve to validate experimental results. As the illustrations demonstrate, they are also used in the practice of good science. I have shown that these strategies can be incorporated quite easily within a Bayesian approach to the philosophy of science, supporting both the view that these are good strategies and also that the Bayesian approach to science is both reasonable and fruitful.

Reasonable belief in an experimental result does not, of course, guarantee its correctness. There may be unknown sources of experimental error or unknown background effects that disturb the observation or measurement. Thus, for example, scattering and energy loss in the source made the early measurements of β-decay spectra incorrect, but it was not unreasonable for physicists to have believed the results were valid, and to have made use of them prior to realizing that there were experimental difficulties. Similarly, the Rustad-Ruby result on the angular correlation in He[6] was incorrect because they did not consider the possibility that decays could occur outside the decay region. This result, quite reasonably, was the major piece of evidence supporting the view that part of the weak interaction was tensor. It was only after extensive work by Wu and Schwarzschild and by Rustad and Ruby, themselves, that the problem was perceived and corrected. When all the experimental checks one can think of have been done, and all the plausible sources of error either eliminated or corrected for, then it is reasonable to believe an experimental result is valid. It would be unreasonable to continue the search for experimental errors or sources of background. Diminishing returns set in. As we have seen, however, if an experimental result is important enough the experiment will be repeated, usually by another technique, providing an additional check on the result.[22]

Sometimes the convergence to an accepted result may take a long time. As O'Conor (1937) pointed out, the beta-decay spectrum of RaE had been measured twenty-five times over a period of thirty years with discordant results. By 1940, however, a consensus had been reached. Does this require that the reasons for the earlier, presumably incorrect, results be known? I think not. It would be nice if they were known, and sometimes they are (recall the discussion of the Rustad-Ruby experiment), but if the

---

[22] See Franklin (1986, ch. 8) for other examples.

arguments for the newer results are strong enough then I believe one can accept them as valid. I will return to this point in Chapter 8 when I discuss the experiments on atomic parity violation.

Life is sometimes difficult on the frontier. This is true for both experiment and theory. Difficulty does not, however, mean impossibility. Science may be fallible, but it is not necessarily wrong. There are, after all, experimental results that we continue to believe are correct.

In the next section I will show how a group of physicists actually applied these strategies in an experiment looking for very rare events.

6.3 MEASUREMENT OF THE K^+_{e2} BRANCHING RATIO: FINDING A NEEDLE IN A HAYSTACK

In this section I will discuss a particular experiment (Bowen et al. 1967) and describe in detail the procedures used to arrive at a final experimental result that could be compared with theoretical predictions.[23] Not too surprisingly, we will see that the strategies discussed in the last section were used. In addition, the experiment chosen, the measurement of the K^+_{e2} branching ratio, was a further test of the *V-A* theory discussed earlier.[24] Even though the theory was already strongly supported by the evidence, as we have seen, there was still good reason to test it under different circumstances and in different phenomena.[25]

The *V-A* theory predicted that the ratio of the decay rates for $(K^+ \rightarrow e^+ + \nu)/(K^+ \rightarrow \mu^+ + \nu)$ would be 2.6×10^{-5} for a pure axial vector, *A*, interaction (without radiative corrections), whereas a pure pseudoscalar interaction, *P*, would give a value of 1.02. Thus, a measurement of the K^+_{e2} branching ratio, the fraction of all K^+ decays that go into a positron and a neutrino,

---

[23] In a recent paper Bogen and Woodward (1988) discuss the distinction between data and phenomena. In their view, theory makes predictions about phenomena not about data. I think that this is a useful distinction although I prefer to use "experimental result" for their term "phenomena." In this section I will show how the group went from data to a result.

[24] I was also a member of that experimental group and was one of two physicists, David Bowen was the other, who had primary responsibility for the data analysis.

[25] This experiment tested the theory in strangeness-changing weak interactions, which had not been previously tested.

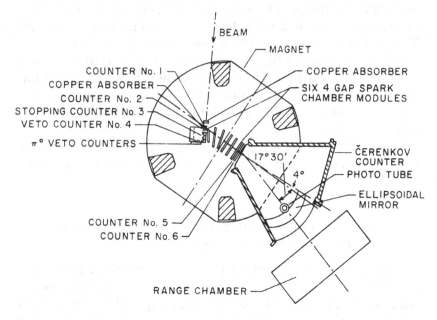

Figure 6.4. Details of the experimental arrangement for the K^+_{e2} experiment of Bowen et al. (1967).

provides a stringent test of the theory and on the presence of the P interaction.

6.3a Experimental apparatus

The experimental apparatus is shown in Figure 6.4. The first order of business was to obtain a supply of stopped K^+ mesons. The reason why they had to be at rest will be discussed later. The group obtained an unseparated positive beam of momentum 530 MeV/c from the Princeton-Pennsylvania Accelerator (PPA). This beam included pions and protons, in addition to the K^+ mesons (kaons) needed. The kaons were identified by their range in matter and by time of flight. The beam telescope consisted of four scintillation counters, C_1, C_2, C_3, and C_4, with 6.7 cm of copper placed before the stopping region, which was counter C_3. A stopped particle was indicated by a coincidence between C_1, C_2, and C_3 with no pulse in C_4 ($C_1C_2C_3\overline{C_4}$). The copper eliminated virtually all of the protons, which have a shorter range in matter than kaons, before they reached C_3. In addition, protons do not decay into positrons.[26] Pions were a more serious problem. There were about 1000 times as many pions as kaons in the beam. Most of the pions, which have a longer range than kaons, passed through the stopping region and counted in C_4, and were eliminated, reducing the ratio of pions to kaons to about 100 to 1. Time of flight provided additional discrimination. Particles of the same momentum (this was a momentum selected beam) but different masses have different velocities and therefore different times of flight. The internal proton beam at the PPA consisted of bunches of protons separated in time by 34 ns. Thus, particles were produced every 34 ns. A signal from the RF (radio frequency) system of the accelerator signaled the production of particles and could therefore be used to time the beam particles. For the beam transport system used in this experiment the difference in time of flight between pions and kaons was 8 ns, so a narrow coincidence of 3 ns ($C_1C_2C_3\overline{C}_4$ + RF) was used to separate kaons from pions. The background of unwanted pions was approximately 5 percent.

During the initial testing of the beam an unexpected and interesting artifact appeared. A typical momentum spectrum for such a beam has a peak at the desired momentum with a reasonable spread. This was observed, but as the momentum of the beam was increased the number of particles began to rise again. This was unexpected and caused some consternation in the experimental group. Fortunately, an explanation was found. The rise in particles occurred at a momentum approximately twice the designed momentum. At such a momentum, protons, which are about twice as heavy as kaons, will have the same speed as the kaons with the designed beam momentum and thus satisfy the time of flight criterion. Protons of this higher momentum will also penetrate the copper absorber and come to rest in C_3, the stopping region. They will look just like the desired kaons. The increase in the number of beam particles was due to protons. In fact, having found the explanation for the unexpected rise, the experimenters strengthened their belief in the proper operation of the apparatus. This was not an artifact known in advance to be present, but

---

[26] Although modern theory does predict proton decay into a positron, both the measured proton lifetime and the predicted rate are far too small to be of any significance in this experiment.

having found the explanation, the experimenters realized that they should have known in advance that it would be there.

Particles from decays at approximately 90° to the incident beam passed through a set of thin plate spark chambers, the momentum chambers, located in a magnetic field. This permitted the measurement of the particle momentum with a resolution of 1.9%. The decay particles were detected by two scintillation counters, C_5 and C_6, and then passed through a gas Cerenkov counter, which was set to detect high energy positrons. This Cerenkov counter had been independently checked, and had a measured efficiency for 250 MeV positrons of between 95 and 99 percent, depending on the trajectory of the positron through the counter. For particles other than electrons, the measured efficiency was approximately 0.38%. Thus, the counter served to identify decay positrons.

A thick plate range spark chamber was placed behind the Cerenkov counter. Its total thickness of 80 g/cm^2 was enough to stop all particles resulting from the decay of the K^+ meson.[27] The time between the K^+ stop and the decay particle was also recorded for each event.

A K^+ decay was identified by a coincidence between a stopping K^+ signal ($C_1C_2C_3\overline{C}_4$ + RF) and a decay particle pulse (C_5C_6). If the events were really due to K^+ decays the time distribution between the K^+ stop and the decay pulse should match the known K^+ lifetime. This experimental check was performed and the positive results found are shown in Figure 6.5. An electronic gate was used to eliminate pion background, as shown in the figure. A K^+ decay into a positron was identified by a decay signal (C_5C_6 + Cerenkov counter).

6.3b Data analysis

Background

Before we begin our discussion of the data analysis it is worth looking at what signal the experimenters expected to find, and

---

[27] Another set of counters, that detected gamma rays, was placed above and below the stopping region. This served to reduce the background from K^+_{e3} decay. This decay was $K^+ \rightarrow e^+ + \pi^0 + \nu$. The π^0 decayed into two gamma rays, which were then detected, and served to eliminate these events.

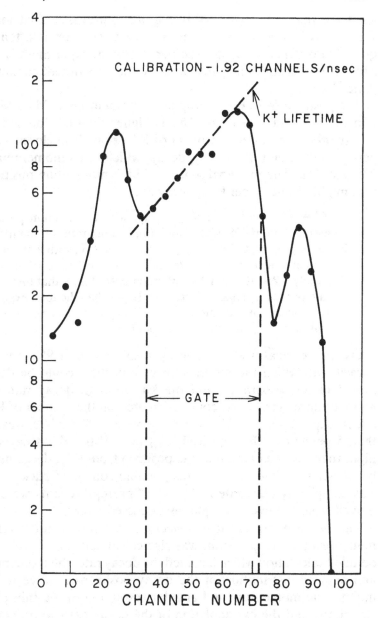

Figure 6.5. Time distribution between a stopping K⁺ pulse and a decay particle pulse. From Bowen et al. (1967).

what the possible sources of background, processes that might simulate K^+_{e2} decay, were. This sort of calculation is often required in order to estimate whether or not an experiment is feasible. If the backgrounds are too large the experiment cannot be done.[28]

The positron from K^+_{e2} decay has a momentum of 246.9 MeV/c in the K^+ center of mass. This is higher than the momentum of any other direct decay product of K^+ decay. The closest competitor is the muon from $K^+_{\mu2}$ decay, which has a momentum of 235.6 MeV/c. The principal sources of high-momentum positrons that might mimic a real K^+_{e2} decay are

1. $K^+\rightarrow e^+ + \pi^0 + \nu$, K^+_{e3} decay, with a maximum positron momentum of 228 MeV/c and a branching ratio of about 5%.
2. $K^+\rightarrow \mu^+ + \nu$, followed by $\mu^+ \rightarrow e^+ + \nu + \bar{\nu}$, with a maximum momentum of 246.9 MeV/c and a branching ratio of approximately 1.2×10^{-4} per foot of muon path. I note that this decay rate per foot is about a factor of 5 larger than the total expected K^+_{e2} decay rate. If this source of background can not be eliminated the experiment can not be done.

Using the measured momentum resolution of 1.9%, the accuracy with which the momentum of a particle could be determined in the apparatus, and the known K^+_{e3} decay rate and momentum spectrum one could calculate that the number of K^+_{e3} events expected in the K^+_{e2} decay region, 242–52 MeV/c, was less than 5 percent of the expected K^+_{e2} rate. If the K^+ decayed in flight then the momentum of the positron from K^+_{e3} decay might be higher than 228 MeV/c. This possible source of background was completely eliminated when the "prompt" events were removed from the event sample, as discussed later.

The background due to K^+ decay into a muon, followed by muon decay into a positron, was also calculated. This involved a detailed calculation which included the decay rate, the momentum and angular distribution of the decay positrons relative to the muons, the momentum and angular resolution in the thin plate chambers, and the extrapolation of the decay particle trajectory from the thin plate chambers into the range chamber, discussed

---

[28] As Peter Galison (1987) has emphasized, and as we will see in detail later, the estimation of and elimination of background is central to the experimental enterprise.

later. The result of this calculation was an expected background of about 15% of the expected K^+_{e2} decay rate. The experimenters expected a total background of approximately 20% of the expected K^+_{e2} rate. There were other sources of background, due to the operation of the experimental apparatus, but these were not calculable in advance, and will be discussed later.

Reduction of data

Figure 6.6 shows the momentum distribution of 16,965 events obtained with the Cerenkov counter in the triggering logic. These events satisfied the following criteria: (1) the track came from the stopping region, C_3; (2) the track passed through both the front and rear windows of the Cerenkov counter; and (3) a track traversing at least three plates was seen in the range chamber. This is the haystack from which the needle of a few K^+_{e2} events was to be found. The momentum for K^+_{e2} decay is shown. It is clear that if the events are there they are rather well hidden.

The large peak at 236 MeV/c and the smaller peak at 205 MeV/c are due to accidental coincidences between accelerator produced background in the Cerenkov counter and muons and pions from $K^+_{\mu2}$ and $K^+_{\pi2}$ decay, respectively.[29] If the K^+_{e2} events are to be found then the background due to $K^+_{\mu2}$ events must be reduced. The fact that the apparatus can detect this muon peak, however, also gives us confidence that it can detect charged decay products. In addition, because the momentum of these muons is known, this peak can be used both to calibrate the momentum scale and to measure the momentum resolution.

We now begin a Sherlock Holmes strategy. In this case we are eliminating the unwanted, rather than the impossible.[30] The first criterion applied was that of range, the path length in the range chamber before the particle stopped or produced an interaction. Muons lose energy only by ionization loss and thus have a well-defined range in matter. The measured range for muons from $K^+_{\mu2}$ decay, selected by the decay particle having a momentum

[29] $K^+_{\mu2}$ decay is $K^+ \rightarrow \mu^+ + \nu$, and $K^+_{\pi2}$ is $K^+ \rightarrow \pi^+\pi^0$.
[30] This might be more appropriately called the Michelangelo strategy. According to an apocryphal story, Michelangelo was asked how he sculpted the statue of David. He replied, "You start with a block of marble and remove everything that doesn't look like David." In this case we try to eliminate all events that don't look like K^+_{e2} decays.

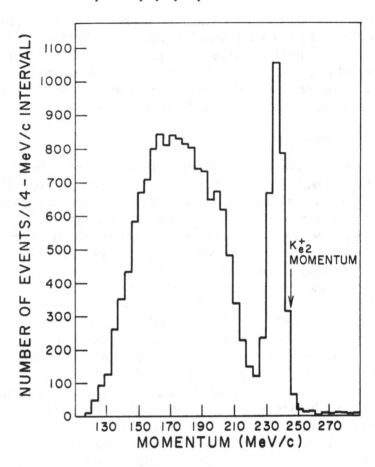

Figure 6.6. Momentum distribution of all K^+ decays obtained with the Cerenkov counter in the triggering logic. The momentum of the K^+_{e2} decay is shown. From Bowen et al. (1967).

between 231 and 241 MeV/c, is shown in Figure 6.7. The muons have a mean measured range of 67 g/cm^2, with a straggle of about 4 g/cm^2. The 1 percent of such events with a range less than 45 g/cm^2 is too large to be accounted for by range straggling, and was due to the occasional failure of the range chamber to operate properly. These apparatus failures gave rise to an expected background of about 15 percent of the expected K^+_{e2} rate. The range for positrons with momenta between 212 and 227 MeV/c is shown in Figure 6.8.[31] These positrons differ from K^+_{e2} positrons by only

---

[31] A small correction was made for pions from $K^+_{\pi2}$ decay.

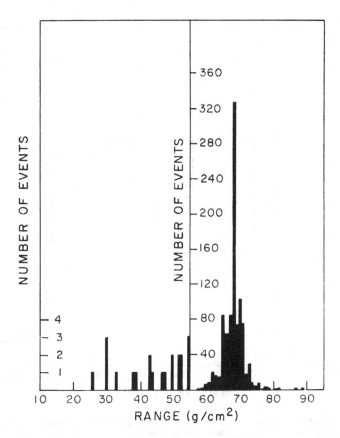

Figure 6.7. Range spectrum for $K^+_{\mu 2}$ muons. From Bowen et al. (1967).

10 percent in momentum, and were expected to behave quite similarly. Positrons do not have a well-defined range because they lose energy by several different processes, some of which involve large energy loss, and the distribution of ranges is approximately constant from about 15 to 70 g/cm². The percentage of positrons with range less than a given value is shown in Table 6.1. If events are required to have a range less than that of the muon from $K^+_{\mu 2}$ decay, this will serve to minimize the background due to those events, while preserving a large, and known, fraction of the high-energy positrons. A selection cut was made at 45 g/cm². The effect of applying this criterion is shown in Figure 6.9. The haystack has gotten smaller.

As discussed earlier, a major source of background was decay

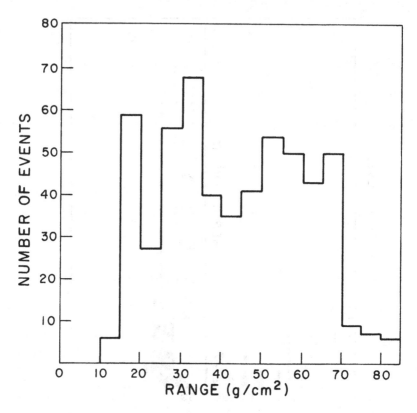

Figure 6.8. Range spectrum for K^+_{e3} positrons with momentum between 212 and 227 MeV/*c*. From Bowen et al. (1967).

Table 6.1: *The range distribution of positrons from* K^+_{e3} *decay in the momentum interval 212–27 MeV/c*

Range (g/cm²)	Percentage of events with smaller range
40	45.6 ± 2.1
45	51.9 ± 2.1
50	59.2 ± 2.1
55	68.8 ± 1.9

Source: From Bowen et al. (1967).

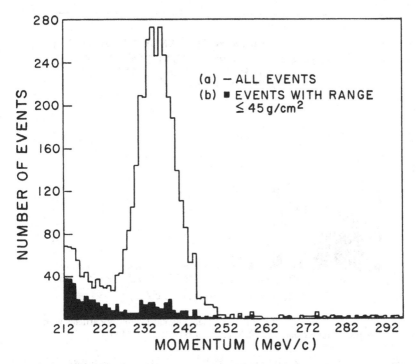

Figure 6.9. The effect of the range cut for decay particles with momentum greater than 212 MeV/c. From Bowen et al. (1967).

of the kaon into a muon, followed by the decay of the muon into a positron. Most of these positrons are emitted at large angles to the muon path. If the decay occurred in the momentum chambers, it would have been detected by a kink in the track. Decays occurring between the end of the momentum chambers and the end of the Cerenkov counter, a very long distance, could not be seen. Because of the large decay angle such decays could be detected by comparing the measured position of the particle when it entered the range chamber with the position predicted by extrapolating the momentum chamber track. If a decay had occurred, then the difference between the two positions would be large. The accuracy of the comparison was limited by multiple scattering in the momentum chambers and the uncertainty in extrapolating the path through the fringe field of the magnet. Such a track-matching criterion was applied.[32] The effects of this selection cut

---

[32] The track-matching criterion required that the measured position of the track

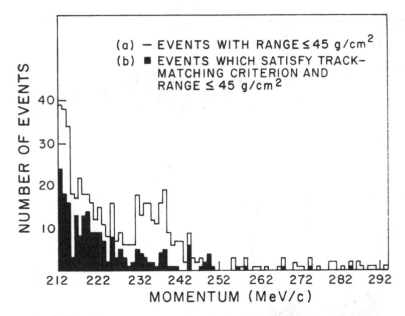

Figure 6.10. The effect of the track-matching cut for decay particles with momentum greater than 212 MeV/*c*. From Bowen et al. (1967).

are shown in Figure 6.10. Once again, the selection criteria served to preferentially reduce the events in the $K^+_{\mu 2}$ region relative to the events in the K^+_{e2} region.

There was one further major source of background. This was due to decays in flight of the K^+ meson. If the kaon decayed in flight then the momentum of the decay particle could be increased, leading to possible simulation of K^+_{e2} decays. Examination of the distribution of time intervals between the stopped kaon and the decay positron revealed the presence of a small peak due to such decays in flight. The peak had a base width of 2 ns. A cut was made removing all events with a time interval of less than 2.75 ns. This eliminated all of the decays in flight. The effect of this selection criterion is shown in Figure 6.11. The effect of this cut was to preferentially reduce the events in the K^+_{e2} region, indi-

agree with the extrapolated position to within − 6 cm to + 10 cm in the horizontal direction and to within − 7 cm to + 6 cm in the vertical direction. The acceptance interval was larger in the horizontal direction because of the greater uncertainty in extrapolating through the fringe field. The center of the distribution varied slightly with momentum, and the fiducial area was adjusted appropriately.

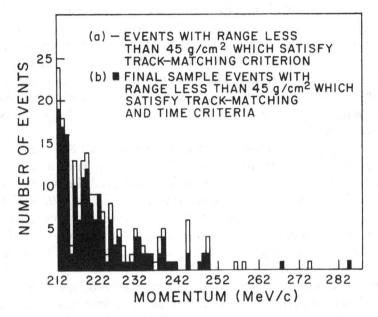

Figure 6.11. The effect of the K^+-decay-time cut for decay particles with momentum greater than 212 MeV/c. The shaded events are the final sample. From Bowen et al. (1967).

cating that decays in flight were indeed a source of simulated K^+_{e2} events.

A total of seven events remained in the K^+_{e2} region. Unfortunately, none of these had a label identifying them as a real K^+_{e2} decay. How one calculated the number of such decays and the branching ratio is discussed in the next section.

The experimental result

Some of the events in the K^+_{e2} region were due to background even after all the selection criteria had been applied. If we examine the region of Figure 6.11 above 252 MeV/c, where no high-momentum positrons from K^+ decay are expected, we find two events in a 30-MeV/c interval. Assuming that this background is constant, there are 0.67 ± 0.5 background events in the K^+_{e2} region, 242–52 MeV/c. The number of events in the K^+_{e2} region must also be corrected for background due to $K^+_{\mu2}$ decay. Analysis of the $K^+_{\mu2}$ region, 231–41 MeV/c, correcting for flat background, K^+_{e3} events, K^+_{e2} events, and momentum resolution, gave an expected background of 2.1 ± 0.5 events in the K^+_{e2} region. Thus, a total of $4.2_{-2.6}^{+3.7}$ remained, and these were attributed to K^+_{e2} decay.[33] Correction for the finite momentum interval used and the momentum resolution gave a final total of $6.0_{-3.7}^{+5.2}$ K^+_{e2} events. The experimenters noted that the probability that all of the original seven events were due to background, when only 2.8 were expected, was 1.6%, an unlikely occurrence.[34]

The branching ratio, the rate compared to all K^+ decays, was calculated by normalizing the K^+_{e2} events to known K^+ decay rates by two different methods. The first used the upper end of the K^+_{e3} spectrum, the region from 212 to 228 MeV/c in Figure 6.11, which had been subjected to the same selection criteria as the K^+_{e2} events. The second method used the total sample of 16,965 K^+ decays given in Figure 6.6. The results for the branching ratio were $R = (2.0_{-1.2}^{+1.8}) \times 10^{-5}$ and $R = (2.2_{-1.4}^{+1.9}) \times 10^{-5}$, respectively. The two different methods, which had very different selection criteria, agreed and the final result given was their average, $R = (2.1_{-1.3}^{+1.8}) \times 10^{-5}$. This was in good agreement with the value, $R = 1.44 \times 10^{-5}$, predicted by the *V-A* theory.[35]

The experiment has been done several times since 1967, with a current total number of K^+_{e2} events of approximately 1000. The currently accepted value for the K^+_{e2} branching ratio is $(1.54 \pm 0.07) \times 10^{-5}$. Within a rather large experimental uncertainty, Bowen and his collaborators had found the needle in the haystack.

6.3c Discussion

In discussing the K^+_{e2} experiment, it should be emphasized that the application of the selection criteria did not guarantee an experimental result in agreement with theory. The experimenters did not stop applying the cuts when a result in agreement with theory was obtained. In fact, when the final sample of seven events was found, the experimenters did not know what the branching

---

[33] For a better estimate of the uncertainty, Poisson statistics rather than Gaussian statistics were used here. This seemed more appropriate for such a small number of events.

[34] This was calculated and reported in the original paper.

[35] This value included radiative corrections.

Table 6.2: *The effect of the track-matching criterion for events with range less than 45 g/cm²*

Momentum region (MeV/c)	Accepted intervals	Accepted intervals increased by 2 cm	Accepted intervals decreased by 2 cm
K^+_{e3} (212–28)	161	177	143
$K^+_{\mu2}$(231–41)	28	35	20
K^+_{e2}(242–52)	13	13	11
$P > 252$	5	6	4

Changes in the accepted intervals were made in both the horizontal and vertical directions. No time cut was made.
Source: From Bowen et al. (1967).

ratio would turn out to be. The observed number of K^+_{e2} events could have been far larger or far smaller than theory predicted. Although, as we have discussed, the experimenters had an estimate of the number of events expected if the theory was correct and had good reasons for believing that the application of these selection criteria would enhance the signal, they could have been wrong.

It should also be emphasized that the experimental result was robust under reasonable changes in the selection criteria. This is illustrated in Table 6.2, which gives the number of events in various momentum intervals for varying track-matching criteria. As the intervals are increased or decreased, the number of events also increases or decreases. The ratios of events in the different regions, however, remain constant, within statistical uncertainty, leaving the final result unchanged. Increasing the range of accepted particles to 55 g/cm² or reasonable changes in the decay time criterion also did not change the result. Eliminating the decay-time criterion completely, allowing prompt events into the sample, increased the value of R to $(3.5 \pm 1.4) \times 10^{-5}$ and $(3.8 \pm 1.6) \times 10^{-5}$, for the two normalization methods, respectively. One expected, on a priori grounds, that including such events would increase the number of background events that simulated K^+_{e2} decay, and thus, the branching ratio. It did, but even then the result was in reasonable, although not good, agreement with both the final experimental value and with theory. Most, if not all, experimenters do check that their results are not critically dependent on the selection criteria used.

As I suggested earlier, the experimenters used some of the strategies discussed in the previous section to validate their result.[36] Thus, for example, the ability of the experimental apparatus to detect K^+ decays was both checked and demonstrated when the distribution of times between the K^+ stop and the decay pulse matched the known K^+ lifetime. The fact that the apparatus could measure momentum reliably was shown with the detection of the muon peak from unwanted $K^+_{\mu 2}$ decays.[37] The known momentum of the muons was used both to calibrate the momentum scale and to measure the momentum resolution, a quantity needed for the calculation of the final result. The ability of the Cerenkov counter to detect electrons of the appropriate momentum efficiently was both demonstrated and measured in an independent experiment. Particles known to be electrons passed through the Cerenkov counter and were detected.

In calculating the final result the experimenters also used two different methods of normalization. The agreement of the two calculations strengthened their confidence in the result. Statistical arguments were also used to argue that not all the K^+_{e2} events were due to background.[38]

This experiment was not selected because it illustrated the use of the strategies discussed earlier. It was selected because David Bowen and I were responsible for most of the data analysis and I had detailed knowledge of how that analysis was done. I believe that if one looks at any physics experiment one will find that some, or all, of these strategies and arguments are used to validate the results. Other examples include the experiments that demonstrated parity nonconservation, that showed *CP* violation, and Millikan's oil-drop experiment (Franklin 1986, ch. 7).

Does the fact that scientists use these strategies argue for their

---

[36] Not all of the strategies were used. I believe that physicists are unscrupulous opportunists and will make use of any strategies they can to argue for the validity of their result. Which strategies can be applied depends, of course, on the particular experimental apparatus.

[37] One might even say that the muon peak was an artifact of the experimental apparatus.

[38] I was not a Bayesian, conscious or otherwise, at the time. I was, however, willing to bet 60 to 1 that we had seen K^+_{e2} decays. I made such a statement in a talk I gave on the experiment, at the time.

legitimacy? After all, not all of the common behavior of scientists has methodological significance. I think so. I believe that science has provided us with reliable knowledge about the world using these procedures.[39] We can generate electricity, build complex particle accelerators, and put a man on the moon, and so forth. This would seem, at the very least, to argue for the pragmatic efficacy of these strategies. I have, in addition, presented arguments for their philosophical validity.

6.4 SOME GENERAL COMMENTS ON EXPERIMENT

The importance of selectivity in an experiment cannot be overemphasized. The K^+_{e2} experiment just discussed was an exercise in such selectivity. As we have seen, both the design of the experiment and the analysis of the data, which selected high-momentum positron decays of the K^+ meson and eliminated background events which might have simulated that process, served to isolate or, at least, enhance observation of the K^+_{e2} decay. This illustrates and supports Peter Galison's view that elimination of background is central to the experimental enterprise (1987, ch. 4). I believe that these characteristics of selectivity and isolation are essential to experiment.

There are, of course, dangers inherent in such selectivity. The very design of the experiment may, unbeknownst to the experimenters themselves, preclude observation of the effect being looked for or of other phenomena of interest. Thus, in the episode on the double scattering of electrons in the 1930s (Franklin 1986, ch. 2), the design of the experiment, which used the scattering of the beam off the front surface of the target (a reflection experiment) rather than having the beam pass through the target foils (a transmission experiment), gave rise to a plural scattering background that masked the expected asymmetry. In this case, the experiment was designed to minimize the expected background due to multiple scattering in the target by using a reflection-type apparatus. The plural scattering background, which is important only in such a reflection experiment and is negligible in a transmission experiment, was totally unexpected.

---

[39] I am not discussing here the question of the existence of theoretical entities or the question of scientific realism.

I believe, however, that such dangers must be risked in order to perform experiments.

I have dealt thus far with selection procedures that appear in the published papers. Much selectivity never appears in the publication but happens during the selection and analysis of data. For example, no experimenter would include in the published paper data taken when they were not sure that the experimental apparatus was working properly. Millikan's notebook for his famous 1913 oil-drop experiment that measured e, the charge on the electron, shows that he excluded the data from his first sixty-eight drops because he was not confident that his apparatus was working properly. Convection in the air was the primary problem. It was only when such effects were reduced that Millikan was willing to accept data for publication (Franklin 1986, ch. 5). He also excluded some of the data taken after he was generally confident of the operation of the apparatus. Most of them were excluded for good reasons such as the failure of the apparatus for a particular drop, or because the experimental conditions required a second-order correction to Stokes's Law, one that was very unreliable, in order to calculate the value of e. Five events seem to have been excluded solely because their values of e did not agree with Millikan's expectations and would have increased his statistical uncertainty. The effect of this exclusion on Millikan's published value of e was quite small, and Millikan's reputation as a careful and accurate experimenter remains.[40] More serious, however, was Millikan's exclusion of a drop which gave the value of the charge on the electron as approximately ⅔ e. Millikan knew this result, wrote, "Won't work," in his notebook, and never published this event. This result seems to have been excluded because it did not agree with Millikan's belief that charge is quantized.[41] Millikan was indeed fortunate that free quarks, or fractional electrical charges, do not appear to exist. His presuppositions might have caused him to miss an important discovery.

Selectivity may also apply to calculations. There are two meth-

---

[40] I note that Millikan won the Nobel Prize for this work.

[41] This may seem to give some support to those who believe that the theoretical presuppositions of an experimenter influence the experimental result. This issue is discussed later. I note, however, that for such an important measurement repetition did, in fact, act as a sufficient safeguard.

ods that can be used to calculate *e* from Millikan's data. Although Millikan claimed that he had used the more reliable method exclusively, he achieved his consistency and reduced his statistical uncertainty by using a combination of methods on some of his events. Once again the effect on the final value of *e* was quite small. (See Franklin 1986, ch. 5, for details.)

Although this invisible selectivity may weaken our belief in the validity of an experimental result, I don't believe that this should be a large effect. Experimenters want to get the correct result and will not, in general, use selection procedures that they know are faulty. I also believe that there are good reasons for virtually all of the selection procedures used by experimenters, and that scientists are aware of such arguments. Consider the discussion of the selection procedures used in the K^+_{e2} experiment given earlier. In addition, repetition of experiments will act as a safeguard.

Some recent work has emphasized the plasticity of experiment and its results. Andy Pickering's (1987) important and provocative essay argues that the theory of the phenomenon under investigation, the experimental apparatus, and the theory of that apparatus are plastic resources that the investigator brings into mutual adjustment. Pickering's example is Morpurgo's search for free quarks with fractional electric charge, using an improved Millikan oil-drop type of apparatus.

Morpurgo began from a conceptual design study of what he believed an adequate charge-measuring device should look like. He set out to implement this design in the material world, to build the apparatus. When the apparatus had been built, he attempted to use it to measure charges (on samples of graphite, initially). And he found that it did not work. Instead of finding integral or fractional charges, he found that his samples appeared to carry charges distributed over a continuum. There followed a period of tinkering, of pragmatic, trial-and-error, material interaction with the apparatus. This came to an end when Morpurgo discovered that if he increased the separation of the capacitor plates within his apparatus he obtained integral charge measurements. . . . After some theoretical analysis, Morpurgo concluded that he now had his apparatus working properly, and reported his failure to find any evidence for fractional charges. . . . Morpurgo would have been happy if his tinkering had eventuated in measurements of fractional charges.

In fact, it did not. The point I want to emphasize is that this eventuation was not entirely under Morpurgo's (or anyone's) control; it was a product of Morpurgo's immersion, through the medium of his experiment, in the real (Pickering 1987, p. 197).

Pickering goes on to note that Morpurgo did not tinker with the two competing theories of the phenomena then on offer, those of integral charge or of fractional charge.

The initial source of doubt about the adequacy of the early stages of the experiment was precisely the fact that their findings – continuously distributed charges – were consonant with neither of the phenomenal models which Morpurgo was prepared to countenance. And what motivated the search for a new instrumental model was Morpurgo's eventual success in producing findings in accordance with one of the phenomenal models he was willing to accept.

The conclusion of Morpurgo's first series of experiments, then, and the production of the observation report which they sustained, was marked by bringing into relations of mutual support of the three elements I have discussed: the material form of the apparatus and the two conceptual models, one instrumental the other phenomenal. Achieving such relations of mutual support is, I suggest, the defining characteristic of the successful experiment (Pickering 1987, p. 199).

Pickering has made several important and valid points about experiment. Most importantly, he has emphasized that an experimental apparatus is rarely initially capable of producing valid experimental results. One of the arts of the experimenter is to modify the apparatus until it does. Recall the improved magnetic shielding required before the experimental apparatus constructed to measure the spin of He^6 was regarded as giving valid measurements. The efforts devoted to getting the K^+_{e2} apparatus to work took more than a month.

He has also recognized that both the theory of the apparatus and the theory of the phenomena can enter into the production of a valid experimental result. What I wish to question is the emphasis he places on these theoretical components. I have already suggested that these theoretical components can be among the strategies used to argue for the validity of an experimental result. I do not believe, as Pickering seems to, that they are necessary parts of such an argument. As Hacking (1983) has pointed out, experimenters had confidence in microscope images both before and after Abbe's work fundamentally changed the theoretical understanding of the microscope. This was due to intervention, not theory.

Pickering seems to neglect the fact that prior to Morpurgo's experiment it was known, or there were at least excellent reasons to believe, that electric charge was quantized in units of *e*, and that fractional charges, if they existed, were very rare in comparison to the integral charges.[42] From Millikan onward, experiments have strongly supported the existence of a fundamental unit of charge, and charge quantization.[43] The failure of Morpurgo's apparatus to produce measurements of integral charge indicated that it was not operating properly and that his theoretical understanding of it was faulty. An analogous case might be a new instrument designed to measure atomic spectra. If, when used on hydrogen, it produced a continuous spectrum, rather than the known line spectrum, one would doubt the validity of its measurements. It was the failure of Morpurgo's apparatus to produce measurements in agreement with what was known, to fail an important experimental check, that caused the doubt about the validity of its measurements. This was true regardless of what theoretical models were available. It was only when Morpurgo's apparatus could reproduce known measurements that it could be trusted and used to search for fractional charges.

Pickering (1984a, 1984b) and Galison (1987) have also raised the question of whether or not the theoretical presuppositions of experimenters influence their experimental results. Such presuppositions may cause an experimenter to exclude data, to overlook unexpected results, and to either overlook or misestimate important sources of experimental uncertainty or background. This is quite worrisome for those, like myself, who believe that experimental evidence plays the primary role in the testing, confirmation, and refutation of theories, and is the major factor in theory choice. I believe that theoretical presuppositions can be another source of fallibility of experimental results, but that the

---

[42] Pickering might be willing to include such empirical evidence as part of Morpurgo's willingness to consider the theoretical models, but he makes no mention of the experimental evidence. In his view, theoretical belief seems to be far more important than experimental evidence.

[43] It is true that Ehrenhaft claimed that charge was not quantized, but there were good reasons to believe that his apparatus did not produce valid results (see Millikan 1916). In addition, the overwhelming preponderance of evidence favors quantization.

corrigibility discussed earlier will also work here. Mistakes may be made, but the history of science shows that they will be corrected.

In his study of gyromagnetic experiments in the early twentieth century, Peter Galison (1987, ch. 2) pointed out that theory often provides experimenters with quantitative predictions that enable them to find the effects sought or to separate these effects from background sources of error. Such predictions may also influence the decision to stop looking for sources of error, declare the experiment ended, and report the result, which may be that predicted by theory. In the early experiments it seems that the theoretical presuppositions of the experimenters did lead to incorrect results, or at least to results that disagree with currently accepted results. They were, however, in agreement with the existing theoretical predictions. In this case, the importance of the experiment led to many repetitions and to the conclusion that the early results were incorrect.

Another case, considered by both Galison (1987, ch. 4) and Pickering (1984b), is that of the experimental discovery of weak neutral currents. Events now attributed to such currents were seen in early experiments, but were thought to be due to neutron background. At the time there was no theoretical prediction of such currents. Later, after theory did predict their existence, such currents were found. The early experiments were also then reinterpreted. It is clear that changes in the evidential context[44] did cause experimenters to search for these events and subsequently to change their interpretation of the earlier results. Such corrigibility is obviously a good thing for science. Theory can tell you what to look for, and which experimental cuts to make or what selection criteria to apply, but it cannot guarantee that the apparatus will produce the expected events.[45] The existence of such events was established using the epistemological strategies discussed earlier. Galison emphasizes the social nature of the decision to end an experiment and report a result that takes place in large, modern, high-energy physics groups. He notes that some

---

[44] This felicitous phrase is due to Trevor Pinch.
[45] During the early 1970s the experimental group I was working with did produce pictures that we now realize were evidence for neutral currents. None of us realized this at the time. The possibility was not even considered, as far as I can recall.

group members are more convinced by some parts of the evidence, and that other group members are convinced by other pieces of evidence. The arguments must, however, as Galison puts it, "stand up in court." As David Cline, one of the experimenters, remarked of the neutral current events, "At present I don't see how to make such effects go away."

Even in a theory-dominated experiment such as the recent discovery of the W boson, in which the Weinberg-Salam unified theory of weak and electromagnetic interactions played such an important role, the experimenters reported a completely new type of event that was not predicted by any existing theory: "We report the observation of five events in which a missing transverse energy larger than 40 GeV is associated with a narrow hadronic jet and of two similar events with a neutral electromagnetic cluster (either one or more closely spaced photons). We cannot find an explanation for such events in terms of backgrounds or within the expectations of the Standard Model" (Arnison et al. 1984, p. 115).[46]

The significant effort devoted to searching for magnetic monopoles, tachyons (faster than light particles), or the current search for the fifth force, a violation of Newton's inverse square law of gravitation, should also give us confidence that results in disagreement with accepted theory are not being overlooked or suppressed. We should also remember that violation of parity and of *CP* conservation, two strongly held symmetry principles, were accepted on the basis of experimental evidence. Among the most important kinds of experiments are those that refute a well-confirmed theory or those that confirm an implausible theory. It is an experimenter's hope to find such unexpected results.[47] In

---

[46] An explanation consistent with the Standard Model was later found for these events. For a fascinating account of this see Taubes (1986). The fact that they were reported even though they could not be explained by the existing well-confirmed and strongly supported theory of the phenomena supports my view that such results will be found and reported.

[47] M. Kreisler, one of the experimenters on tachyons, remarked, "Every researcher hopes that his next experiment will yield a fundamental discovery – a discovery that will be considered not only essential to his particular sub-specialty, but one that will radically change our view of the study of physics. . . . The desire to participate in such a discovery is partially responsible for the large number of experiments testing the predictions of fundamental laws, some of which have been found to fail when tested extemely carefully. It is a healthy sign for the study of physics that there are no 'sacred cows'; if there

the case of the experimental anomalies for the *V-A* theory, I believe that the history shows that the theoretical presuppositions of the experimenters were not an important factor in resolving the anomalies.

Nevertheless, it is a real worry that the hopes and expectations of the experimenter may influence the experimental result. Experimenters do sometimes find what they are looking for, even when the evidence is lacking. Thus, Blondlot and others found evidence for the existence of a new type of radiation, N rays, which is not believed to exist (Nye 1980). This is not merely the fallibility of experiment but rather a case of self-delusion. The issue was resolved when R. W. Wood, an American physicist, visited Blondlot's laboratory and found that the observations of the new rays persisted even when he removed what Blondlot regarded as an essential part of the apparatus. There was a similar resolution of the Cambridge-Vienna controversy on artificial disintegration during the 1920s (Stuewer 1985). Chadwick visited the Vienna laboratory and found that their results, which were in disagreement with those of the Cambridge group, also persisted after the radioactive source was removed. Here too we have a case of self-delusion. The human scintillation counters observed what they expected and wanted to observe. I suggest, however, that in both of these cases repetition of important results provided an adequate safeguard.[48] The strategy of intervention was also used here. In this case an essential part of the apparatus was removed. One then expected the effect to disappear. The persistence of the observed effect argued against its validity.

A related issue is that of the theory-ladenness of observation. Experiments are laden with both the theory of the apparatus and with the theory of the phenomenon under investigation. The ques-

is any reasonable chance that something new exists, researchers will spare no effort in the search" (Kreisler 1975, p. 429). A Bayesian analysis can help clarify the importance of such experiments. An experimental result that is entailed by an implausible theory has a very low prior probability. $P(h|e) = P(h)/P(e)$ will therefore be large. A refutation will reduce the probability of the hypothesis to 0, or show the theory is false. There are important questions about this, and these will be discussed in the next chapter.

[48] Although experimenters in France reproduced Blondlot's results, experimenters in England did not. The need to resolve the discrepancy between the two different sets of experiments led to Wood's visit.

tion then arises as to whether or not the experiment can test the theory of the phenomenon in any meaningful way.

If the two theories are different, then there are no apparent problems. One establishes the validity of the experimental results using the strategies discussed earlier. These results can then be used to test theories, to choose between theories, to call for new theories, and in other ways, as discussed in the next chapter.

If the theory of the apparatus is the same as the theory of the phenomenon, then one might worry about circularity. One might have to assume the theory under test is correct in order to perform the experiment. Thus, one might not wish to use a mercury thermometer to test the hypothesis of whether or not objects expand when their temperature increases. The proper operation of the thermometer depends on the truth of the hypothesis being tested. Nevertheless, I believe that even in this case the mercury thermometer could be used. One can establish the validity of measurements in many ways. One could, for example, calibrate the mercury thermometer against another thermometer such as the constant volume gas thermometer, whose proper operation does not depend on the hypothesis.[49] (For further details see Franklin et al. 1989.)

One might also worry about the effects of previous experiments on an experimental result, a kind of experimental bandwagon. One need only look at a survey of the measurements of fundamental physical constants (Cohen and DuMond 1965) or at the "Review of Particle Properties" (Particle Data Group 1986), a standard reference for high-energy physicists, to see that the measured values of not only particular quantities, but also of the world average of such quantities, change by far more than their cited experimental uncertainties would suggest is plausible or probable. Is "the old joke about the experimenter who fights the systematics until he or she gets the 'right' answer (read 'agrees with previous experiments')" (Particle Data Group 1980, p. S286) true? A case in point is the history of measurements of η_{+-}, the *CP*-violating parameter in K^o decay (see Figure 6.12). Prior to 1973, η_{+-} had

---

[49] The reader may object that in this case one should merely use the constant-volume gas thermometer. It may, however, be the case that using a calibrated mercury thermometer is more convenient.

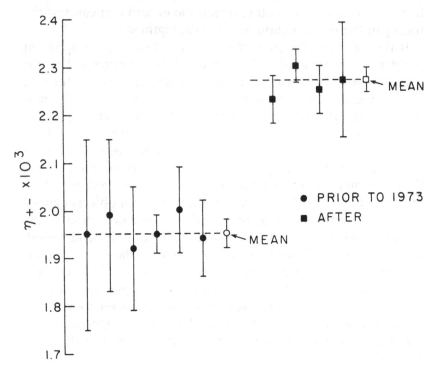

Figure 6.12. Measurements of η_{+-} in the order of their publication.

been measured six times. The results were in good statistical agreement and had a mean of $(1.95 \pm 0.03) \times 10^{-3}$. The four succeeding measurements agreed with each other, and had a mean of $(2.27 \pm 0.022) \times 10^{-3}$. These means differ by eight standard deviations, an unlikely change if these are two sets of correct measurements of the same quantity. (The probability is 1.24×10^{-15}.) Unless one is willing to consider the possibility that η_{+-} changed in 1973, we must conclude that at least one of these sets of measurements is incorrect.[50] This illustrates both the possible problem and its solution. Quantities of interest tend to be mea-

[50] There is still no resolution of the discrepancy. Most physicists accept the later results as correct. "These [the experiments before 1973] are excluded from the . . . average . . . since they do not agree with more recent precise and in principle superior experiments" (Particle Data Group 1986, p. 139). There is, however, no explanation of why the earlier results were wrong. At the moment there is no theoretical calculation of the value of this parameter. Such a prediction might spur the resolution of the problem.

sured several times, usually by different methods. (For other examples see Franklin 1986, ch. 8.)

I believe that these questions of the fallibility of experimental results and of the interaction between theory and experimental results are both important and worrisome. I believe, however, as the examples illustrate, that they are problems that can be overcome. In this chapter I have argued that there are good reasons for believing in experimental results, even though those results are fallible. In the next chapter I will discuss the ways in which those experimental results are used, that is, the roles of experiment.

7

The roles of experiment

7.1 A LIFE OF ITS OWN

After the discussion of the previous six chapters, a reader might be strongly tempted to believe that experiment derives its meaning and significance solely from its relation to theory. We have mentioned the roles that experiment plays in confirming, refuting, and choosing between theories. We have also discussed the role of theory in the validation of experimental results.

As Ian Hacking (1983) has pointed out, however, experiment often has a life of its own. Although we began our history of experiment and the theory of weak interactions with Fermi's (1934) theory of β decay, the subject had been studied experimentally for more than thirty years in the absence of any accepted theory of the phenomena. Similarly, from the discovery of superconductivity by Kamerlingh Omnes in 1911 until the phenomenological theory of London and London in 1935 the experimental study of the effect proceeded in the absence of any successful theory. In fact, one might argue that the experimental discovery of the Meissner effect, the exclusion of a magnetic field from the interior of a superconductor, was a crucial step for the development of the theory. This suggests another role for experiment, that of giving hints toward a successful theory.[1]

Sometimes experiments are done because the phenomena are seen to be interesting. In addition one might wish to acquire data that a future theory will have to explain. This might be the reason for the large amount of effort devoted to measuring atomic spectra

---

[1] How this might happen is beyond the scope of this study, but I suggest that the Meissner effect showed the importance of the magnetic field in superconductivity. The interaction of experiment and theory will be discussed in the next section.

in the nineteenth and early twentieth centuries, or to the discovery and measurement of the properties of strongly interacting elementary particles in the 1960s and 1970s, before any theory could either explain or predict the data. This idea of the future utility of experimental results will also be illustrated in the history of atomic parity violation experiments, discussed in detail in the next chapter. In this case the continuation of the experiments, after the theory of interest had been confirmed, resulted in stringent tests of the theory, which were not envisioned when that type of experiment was begun. In addition, the experiments increased the precision and accuracy of the measurements. This idea of improving the measurement of a physical quantity "because it is there" is not unusual. Witness the large amount of effort devoted to measuring the values of fundamental constants such as e, the charge on the electron, or c, the speed of light, even though there is no current theoretical calculation of these values. As Jack Steinberger, one of the group leaders of the second set of experiments to measure η_{+-}, the CP-violating parameter, remarked, "There was another purely experimental [reason]: we saw a way of doing a much better measurement than had been done" (private communication).

Experimenters also have a major commitment to their instruments or experimental apparatus. This often involves the particular apparatus used. The apparatus used to measure the K^+_{e2} branching ratio had previously been used to measure the major K^+-decay branching ratios. The modifications were small, including the addition of the π° counters and the requirement of the Cerenkov counter in the event trigger (the signal that signified an event of the appropriate type and triggered the apparatus). The subsequent experiment performed by the group measured the π° energy spectrum in K^+_{e3} decay. In this experiment spark chambers were added to measure the directions of the γ rays from the decay of the π° and the event trigger was modified slightly.[2]

The commitment may extend beyond a particular experimental apparatus to what Galison (1987) has called an experimental tradition. In an experimental tradition scientists develop skill in using

---

[2] This experience using the apparatus no doubt increased the experimenters' own confidence in the validity of their results, but other arguments were required to convince, or to persuade, the scientific community.

certain types of instrument and apparatus and will therefore regard particular kinds of evidence as most convincing.[3] In particle physics, Galison discusses the traditions of visual detectors, such as the cloud chamber and the bubble chamber, in contrast to the electronic or logic tradition, which includes Geiger counters, scintillation counters, and spark chambers. Scientists within the visual tradition tend to prefer "golden events" that clearly demonstrate the existence of a phenomenon, such as Anderson's photograph of the positron. In the electronic tradition, statistics tend to be more important than single events. The K^+_{e2} experiment is an example of such an electronic experiment.

Both Galison and Ackermann (1985) argue that experimental practice and instruments and the data they provide persist across major changes in theory, and provide continuity across these conceptual changes.[4] The experiments on the gyromagnetic ratio of the electron spanned classical electromagnetism, Bohr's old quantum theory, and the new quantum mechanics of Heisenberg and Schrodinger (Galison 1987, ch. 2). This experiment did indeed have a life of its own.

7.2 EXPERIMENT AND THEORY

Experiment does, however, often take its significance and importance from its relation to theory. We have seen many different interactions between experiment and theory in the earlier chapters. A result such as the "θ–τ puzzle" called for a new theory because it was inexplicable on the basis of existing, accepted theory. The experiments on the β decay of oriented nuclei (Wu et al. 1957) and on the asymmetry in π–μ–e decay (Friedman and Telegdi 1957a, 1957b; Garwin, Lederman, and Weinrich 1957) were crucial experiments. They refuted the hypothesis of parity conservation, supported the hypothesis of parity nonconservation, and called for a new theory of weak interactions.

Experiment may also confirm, or give support to, an existing theory. The experiments on the β-decay spectra and on the approximate constancy of $F\tau_0$ initially supported Fermi's theory. More detailed examination of the results, however, were regarded

---

[3] Scientists can and do change traditions.
[4] The interpretation of the experimental results may change, but at the empirical level there is continuity.

as refuting Fermi's theory and supporting the Konopinski-Uhlenbeck modification of that theory. Experiment may also help to articulate a theory. The experiments on the angular correlation in β decay helped to choose the specific form of the decay interaction. The discovery of the Meissner effect seemed to offer a hint toward the development of the theory of superconductivity by stressing the importance of the magnetic field in the phenomenon.

In all of these roles we must consider the possibility that the experimental results are incorrect. The fallibility and corrigibility of experimental results necessarily extends to the confirmation or refutation of theories or hypotheses based on those results. This is nicely illustrated by the history of nuclear β decay during the 1930s. (In this section I will summarize the history. For details see Chapter 1.)

In 1934, Fermi proposed a theory of such decays. His theory agreed with the shape of the observed energy spectra. It also predicted that the quantity $F\tau_0$ would be approximately constant for different types of decay,[5] where F is the integral of the energy distribution and τ_0 is the lifetime of the transition. This was confirmed by Sargent's previously reported work (1933).

It was quickly pointed out by Konopinski and Uhlenbeck (1935) that more detailed examination of the spectra showed discrepancies. Fermi's theory predicted too few low-energy electrons and an average decay energy that was too high. They proposed a modification of the theory that remedied these deficiencies. They cited as support for their model the spectra of P^{30}, obtained by Ellis and Henderson (1934), and of RaE, found by Sargent (1932). The experimental spectrum of RaE, along with the predictions of the two competing theories, is shown in Figure 1.2. The superiority of the K-U modification is apparent. The K-U model also predicted the constancy of $F\tau_0$. In a 1936 review article that remained a standard reference and was used as a student text into the 1950s, Bethe and Bacher remarked, "We shall therefore ac-

---

[5] Fermi considered "allowed" decays, those for which the electron and neutrino wavefunctions could be considered constant over nuclear dimensions. "Forbidden" transitions also exist, and their rate is very much reduced. The shape of the decay spectra for such transitions differs from that of "allowed" transitions, a point we shall return to later.

cept the Konopinski-Uhlenbeck theory as the basis of future discussions" (Bethe and Bacher 1936, p. 192).

Further experimental work during the late 1930s and early 1940s showed that the early energy spectra, as well as others that had subsequently provided support for the K-U theory, had systematic experimental problems. Scattering and energy losses in thick sources had resulted in observed spectra with too many low-energy electrons and with too low an average energy. As sources were made thinner the results approached the predictions of Fermi's original theory (see Figure 1.4). The correct theory should give a good fit to a straight line.

It was also pointed out that the spectrum calculated by Fermi was for "allowed" transitions, whereas RaE and several of the other elements whose beta decay spectra had provided support for the K-U theory were "forbidden" transitions. This had been mentioned earlier, but not much attention had been paid to it. The predicted spectra for the two types of transition were not expected to be the same. When the Fermi theory spectrum for forbidden transitions was calculated, ironically, by Konopinski and Uhlenbeck (1941), the discrepancy disappeared. The spectrum of phosphorus, P^{32}, is shown in Figure 1.5 (from Lawson 1939). The superior fit of the Fermi theory is clear.

The early spectra confirmed the K-U theory and refuted Fermi's theory. After these experimental results were found to be in error, the new results reversed that decision. In 1943, Konopinski published a comprehensive review of β decay and concluded, "Thus, the evidence of the spectra, which had previously comprised the sole support for the K-U theory now definitely fails to support it" (Konopinski 1943, p. 218).

We have seen the fallibility and corrigibility of experimental results, of theoretical comparison with data, and of the confirmation and refutation of theories.

Does this episode, along with the general fallibility of experimental results discussed earlier, argue against a legitimate role for experiment in theory choice or in the confirmation and refutation of theories? I think not. Not all incorrect experimental outcomes result in incorrect theory choices. Thus, both sets of measurements of η_{+-}, discussed in the last chapter, confirm the hypothesis of *CP* violation, which requires only that $\eta_{+-} \neq 0$. The fact that Millikan's value of *e*, the charge on the electron,

$(4.774 \pm 0.009) \times 10^{-10}$ esu, disagrees with the currently accepted value of $(4.803242 \pm 0.000014) \times 10^{-10}$ esu has not changed the support for charge quantization.[6] The early history of β decay shows only that we can be mistaken in our theory choice, and not that there weren't good reasons for that choice.

I have already argued that there are good reasons for reasonable belief in experimental results. These results then provide the basis for theory choice. The fact that reasonable belief does not guarantee truth should not preclude justified and reasonable theory choice.

Some sociologists of science, as well as some philosophers and historians of science, have argued against such a role for experiment. A typical view is given by MacKenzie.

Recent sociology of science, following sympathetic tendencies in the history and philosophy of science, has shown that no experiment, or set of experiments however large, can on its own compel resolution of a point of controversy, or, more generally, acceptance of a particular fact. A sufficiently determined critic can always find reason to dispute any alleged "result." If the point at issue is, say, the validity of a particular theoretical claim, those who wish to contest an experimental proof or disproof of the claim can always point to the multitude of auxiliary hypotheses (for example about the operation of instruments) involved in drawing deductions from a given theoretical statement to a particular experimental situation or situations. One of these auxiliary hypotheses may be faulty, critics can argue, rather than the theoretical claim apparently being tested. Further, the validity of the experimental procedure can also be attacked in many ways (MacKenzie 1989, p. 412).

There are two points at issue here. The first involves the meaning one assigns to "compel." If one reads it, as MacKenzie seems to, as "entail" then I agree that no finite set of confirming instances can entail a universal statement. No matter how many white swans one sees it does not entail that "all swans are white." A single instance can, however, refute a universal statement. Observation of a single black swan refutes "all swans are white."[7] Less artificial examples will be discussed later. It seems to me, however, that a more plausible meaning for "compel" is having

---

[6] The difference between Millikan's value and the modern one is primarily due to different values for the viscosity of air.

[7] This assumes, of course, that whiteness is not a defining characteristic of being a swan. Note the presence of black Australian swans at Leeds Castle.

good reasons for belief. As the episodes discussed later demonstrate, this is the meaning used in scientific discourse.

The second point is a logical one, known to philosophers of science as the Duhem-Quine problem (see Harding 1976), and discussed by MacKenzie. In the usual *modus tollens* if a hypothesis *h* entails an experimental result *e* then ¬*e* (not *e*) entails ¬*h*. As Duhem and Quine pointed out it is not just *h* that entails *e* but rather *h* and *b*, where *b* includes background knowledge and auxiliary hypotheses. Thus, ¬*e* entails ¬*h* or ¬*b* and we don't know where to place the blame. One can, of course, also question the experimental result ¬*e*.[8]

As discussed later, there is at least one example taken from the practice of science, where if one accepts the experimental evidence, one can solve the problem of where to place the blame. More important, however, is the fact that scientists never confront all the logically possible explanations of a given result. There is usually only a reasonable number of plausible or physically interesting alternatives on offer.[9] In this case scientists evaluate the cost of accepting one of these alternatives in the light of all existing evidence. One of these alternatives may be far better supported by the evidence than any of the others. The alternatives themselves may also be tested, subject to the usual difficulties, and we may be left with only one explanation. An example of this will also be presented later.

Are there any hints that one might offer as to how one goes about solving the problem of where to place the blame? Noretta

---

[8] Bob Ackermann (1989) has suggested that the Duhem-Quine problem is a rather old-fashioned philosophy of science and that one should rather frame the discussion in Hacking's terms "speculation, calculation, and experimentation." I agree with him, as will be discussed in detail later, that science is not framed in terms of logical articulation of theory and observation. Nevertheless, I believe that a consideration of the problem, not in terms of logically possible alternatives, but in terms of physically interesting or plausible hypotheses, is of value.

[9] Ackermann (1989) has also suggested that I need a theory of plausibility to complete this discussion. Although there is a normative aspect of my discussion, this discussion refers primarily to the judgments of the scientific community in the actual practice of science. I believe that different scientists will make very different judgments about plausibility and that this is likely to be a good thing. I don't know how scientists restrict the hypothesis space they are considering nor do I have any good suggestions as to how they should do so. For further interesting discussion of this see Langley et al. (1987).

Koertge (1978) has made some very useful suggestions. She suggests two strategies.

1. Check the most accessible source of trouble first.[10] In other words, check those alternative explanations that are most easily tested.
2. Check the most probable source of trouble early on.[11]

As we shall see, these are, in fact, the strategies scientists tend to use. She goes on to discuss the appraisals that go into a solution, *X*, of the Duhem-Quine problem.

1. "How interesting or informative or explanatory would *X* be if it were true?
2. What is the probability that *X* is true?" (Koertge 1978, p. 263).

This will, of course, involve appraisals of the plausibility or probability of the alternatives.[12] It will also involve estimates of the scientific interest of the alternatives. As Koertge points out, it is here that the most serious differences of opinion will occur within the scientific community. Even if one had agreed-upon and clear measures of content, simplicity, depth, heuristic power, and so forth, different scientists might well give these criteria different weights.[13] These criteria may also be applied in different ways in different situations.

Should the fact that there is no prescriptive algorithm for the solution of the problem worry us excessively? I don't think so. It is precisely because of these differing judgments that more of the alternatives are likely to be explored. Sometimes, even a very implausible hypothesis turns out to be correct.[14]

Let us now examine some episodes from the history of twentieth century physics to see if the kinds of solutions advocated by Koertge and myself are adequate descriptions of what has happened.

---

[10] Koertge's discussion is influenced, in part, by *Zen and the Art of Motorcycle Maintenance* (Pirsig 1974). In case of ignition failure test the spark plugs before dismantling the carburetor.
[11] If your carburetor has a history of trouble, check it early on.
[12] A Bayesian approach can be applied here, and Koertge does offer a Bayesian analysis of plausibility appraisals (1978, p. 264–5).
[13] A Bayesian approach allows one to take these differing weights into account by varying the prior probabilities.
[14] Recall the earlier estimates of the prior probability of parity nonconservation.

I begin with the discovery of parity nonconservation, or the violation of space reflection symmetry in nature. (For details see Chapter 4.) The origin of the problem was the "θ–τ puzzle." On one set of criteria, that of identical masses and lifetimes, the θ and τ appeared to be the same particle. However, the θ decayed into two pions whereas the τ decayed into three pions, giving rise to states of different spin and parity. On these criteria the θ and τ were different particles. Lee and Yang (1956) realized that a possible solution to the puzzle would be the nonconservation of parity in the weak interactions. If parity weren't conserved then the θ and τ would merely be different decay modes of the same particle. As we have seen, this was regarded as highly unlikely by some very distinguished physicists.

Alternative solutions to the puzzle were also offered. These included:

1. The heavier of the two particles decayed into the lighter with the emission of γ rays. This was tested by Alvarez, who found no such γ rays.
2. All strange particles[15] come in parity doublets, that is, particles of identical mass and lifetime but with opposite parity. Thus, in addition to the θ and τ, there would be a doublet consisting of the Λ_1 and Λ_2. No evidence for such particles was seen.
3. The spin of the θ and τ was high. Although this was compatible with the evidence for a time, subsequent experimental work showed that it was extremely unlikely.
4. "The most attractive way out is the *nonsensical* [emphasis added] idea that perhaps a particle is emitted which has no mass, charge, and energy-momentum, but only carries away some strange space-time transformation properties" (C. N. Yang in Ballam et al. 1957, ch. 8, pp. 27–8).
5. The π^o emitted in θ decay is a "nearly real" π^o.

The first three cases were regarded as both the most probable and the most easily tested. The tests were performed and the hypotheses eliminated. The last two were regarded as so unlikely that they were not tested before the direct experimental tests of

---

[15] Strangeness was a quantity attributed to some elementary particles to explain some of their peculiar properties. Its existence had been experimentally confirmed. For details see Franklin (1986, ch. 1).

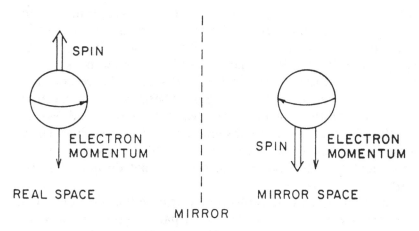

Figure 7.1. The β decay of oriented nuclei in real and mirror space.

parity nonconservation were performed.[16] Although this is not a solution to a particular instance of the Duhem-Quine problem it does indicate that the scientific community makes the kinds of judgments just discussed.

Let us now discuss the direct tests of parity nonconservation. I will consider a weak form of the Duhem-Quine problem, in which one accepts the experimental results as correct. This episode illustrates that there is at least one philosophically legitimate solution to this problem. In this case, there are in fact only two classes of theories, those that conserve parity and those that do not. These classes are mutually exclusive and exhaustive. The experimental test was as follows. Suppose we orient the spin of a radioactive nucleus so that it points upward. Also suppose that the electron from the β decay of this nucleus is emitted in the direction opposite to the spin, or downward. As shown in Figure 7.1, the mirror image of this result differs from the original. In the mirror the electron is emitted along the spin direction. This was the experiment performed by Wu et al. (1957). The results are shown in Figure 4.1 and clearly show an asymmetry.[17] Once

---

[16] The positive outcome of these direct tests made testing of these hypotheses moot.

[17] Similar experiments on the asymmetry in π–μ–e decay were done by Garwin, Lederman, and Weinrich (1957) and by Friedman and Telegdi (1957a, 1957b). The results of Garwin et al. are shown in Figure 4.2. The asymmetry is clearly shown. If there were no asymmetry, the curve would be a horizontal straight line.

one accepts the evidence in favor of the observed asymmetry then it seems that one must either attribute it directly to nonconservation of parity in the weak interactions or else say that parity is conserved in the weak interactions and modify the background beliefs so that the universe itself is asymmetric. There does not seem to be any alternative that preserves both parity conservation and the symmetry of the universe, given the experimental results. It is the very general nature of symmetry principles such as parity conservation that allows them to be refuted. In this case we are compelled to accept the class of theories that do not conserve parity.

A more complex case is the demonstration that the combined symmetry *CP* (charge-conjugation, or particle–antiparticle symmetry, and parity) was violated (Franklin 1986, ch. 3). *CP* conservation allowed the K°_1 meson, but not the K°_2 meson, to decay into two pions. The experiment of Christenson and associates (1964) found $K^\circ_2 \to \pi^+\pi^-$ and thus, in the most obvious interpretation, demonstrated CP violation. This was the view taken in three out of four theoretical papers written during the period immediately following the report of those results. The rest offered alternative explanations. These relied on one or more of three arguments: (1) the Princeton results are caused by a *CP* asymmetry (the local preponderance of matter over antimatter) in the environment of the experiment, (2) $K^\circ_2 \to \pi^+\pi^-$ does not necessarily imply *CP* violation, and (3) the Princeton observations did not arise from $K^\circ_2 \to \pi^+\pi^-$ decay. This last argument can be divided into the assertions that (3a) the decaying particle was not a K°_2, (3b) the decay products were not π mesons, and (3c) another unobserved particle came off in the decay. Included in these alternatives were three suggestions that cast doubt on well-corroborated fundamental assumptions of modern physics. These were: (1) pions are not bosons, (2) the principle of superposition in quantum mechanics is violated, and (3) the exponential decay law fails. Although by the end of 1967 all of these alternatives had been experimentally tested and found wanting, the vast majority of the physics community had accepted *CP* violation by the end of 1965, even though all the tests had not yet been completed. As Prentki remarked, this was because in some cases "the price one has to pay in order to save *CP* becomes extremely high," and because other alternatives were "even more unpleasant"

(Prentki 1966). Prentki is discussing both the probability of the remaining hypotheses and their interest for physicists. This is an example of what one might call a pragmatic solution to the Duhem-Quine problem. Here the alternative explanations, which modified the background knowledge, and the auxiliary hypotheses were refuted, leaving *CP* unprotected. There were certainly other possible alternative explanations, but none were offered, no doubt because they were not regarded as plausible or physically interesting.[18]

Let us examine the experimental tests of these alternative hypotheses in some detail. The first proposal was a cosmological model based on the local preponderance of matter over antimatter (Bell and Perring 1964; Bernstein, Cabibbo, and Lee 1964). They proposed a long-range interaction of cosmological origin that interacted differently with the K^o and $\overline{K}^o$ mesons and thus would regenerate K^o_1 mesons, each of which would then decay into two pions.[19] The rate of such decays would be proportional to the square of the K^o energy. It is not clear how seriously, or how probable, the authors thought this model was. Bell and Perring remarked, "Before a more mundane explanation is found it is amusing to speculate that it might be a local effect due to the dyssymmetry of the environment, namely the local preponderance of matter over antimatter" (Bell and Perring 1964, p. 348). Both Lee and Cabibbo proposed other alternative solutions to *explain* *CP* violation, and Lee proposed still another model to avoid it. There were also theoretical objections to this model (Weinberg 1964; Lyuboshitz, Okonov, and Podgoretskii 1965). Nevertheless, the model used the real asymmetry between matter and antimatter and was analogous to the regeneration of K^o mesons that had previously been observed. This increased its plausibility. In addition, the test of the model using the measurement of the two-pion-decay rate at different energies was likely to have been done because of the smallness of the effect.[20]

---

[18] As we shall see, plausibility was given a rather broad meaning when offering such an alternative explanation. One of them involved a "shadow universe" which interacted with our universe only through the weak interactions. This alternative was not only testable, but was tested.

[19] The phenomenon of regeneration was unique to the K^o meson system and depended on the fact that the K^o and $\overline{K}^o$ interacted differently with matter. For details see Franklin (1986, ch. 3).

[20] The rate of two-pion decay was about 2×10^{-3}. Even though the result of

Two experiments measured the $K^\circ_2 \to \pi^+\pi^-$ at energies different from each other and from that of the original Princeton experiment. DeBouard et al. (1965) and Galbraith et al. (1965) found $(2.24 \pm 0.23) \times 10^{-3}$ and $(2.08 \pm 0.35) \times 10^{-3}$, respectively. The theoretical predictions, based on the Princeton result and the presumed energy dependence, were $(1.6 \pm 0.3) \times 10^{-1}$ and $(13.4 \pm 3.0) \times 10^{-3}$. These results, in agreement with the Princeton value of $(2.0 \pm 0.4) \times 10^{-3}$, eliminated the cosmological model, a point made explicitly by both groups. The results also eliminated other models based on external fields.

Another alternative was that an unobserved particle was emitted in the decay (Levy and Nauenberg 1964; Kalitzin 1965). This was similar to both the original neutrino hypothesis, proposed to save energy and momentum conservation in β decay (a hypothesis that turned out to be correct) and to the Lee-Orear suggestion concerning the decay of the θ into the τ. Both models, which differed slightly from one another, fell when Fitch and his collaborators (1965) observed interference between $K^\circ_2 \to 2\pi$ and $K^\circ_1 \to 2\pi$, where the K°_1 were produced by coherent regeneration from K°_2. As Fitch noted, such interference would be impossible if another particle were emitted in the decay. This interference also ruled out the class of hypotheses that attributed the decay to a particle other than the K°_2.

Cvijanovich, Jeannet, and Sudarshan (1965) suggested that one of the pions supposedly emitted in $K^\circ_2 \to \pi^+\pi^-$ decay is not a pion but rather a particle with the same mass but having spin 1, rather than 0, the so-called spion. This odd explanation had originally been offered to explain a small asymmetry observed in $\pi \to \mu$ decay, particularly in low-energy pions emitted in $K \to 3\pi$ decays (Cvijanovich and Jeannet 1964). The spion would decay by both $\pi \to \mu$ and $\pi \to e$, with roughly equal rates, as opposed to the ratio of 10^4 for normal pion decays. If a neutral spion existed, it would have a preferred decay mode of $e^+e^-\gamma$, with a very short lifetime. Rinaudo et al. (1965) searched for such decays and found no evidence for either effect. Taylor et al. (1965) also examined $K^+ \to 3\pi$ decays and found not only no asymmetry in

Christenson et al. (1964) was statistically significant, 45 ± 9 events, it was still a very small effect.

the $\pi \to \mu$ decay but also no evidence of $\pi \to$ e decays. In addition, the spion model did not allow the interference observed by Fitch. Both Taylor and Rinaudo explicitly mentioned the test of the spion model as an alternative explanation of apparent *CP* violation.

The model of Nishijima and Saffouri (1965) explained $K^{\circ}_2 \to 2\pi$ decay by the existence of a "shadow" universe in touch with our "real" universe through the weak interactions. This model was ruled out by the interference experiments. It nonetheless came in for consideration. Everett (1965) noted that if the $K^{\circ\prime}$ postulated by Nishijima and Saffouri existed, then a shadow pion, or $\pi^{\circ\prime}$, should also exist, and the decays $K^+ \to \pi^+\pi^{\circ}$ and $K^+ \to \pi^+\pi^{\circ\prime}$ should occur with equal rates. Although the $\pi^{\circ\prime}$ could not interact with any apparatus in our universe, its presence could be detected by measuring the $K^+ \to \pi^+\pi^{\circ}$ branching ratio in two different experiments, one in which the π° was detected and one in which it was not. If the $\pi^{\circ\prime}$ existed, the two measurements would differ. Everett examined the evidence available at the time and could draw no firm conclusion. Callahan and Cline (1965) remeasured the $K_{\pi 2}$ branching ratio, detecting only the π^+, and found it to be 21.0 $\pm$ 0.56 percent, in good agreement with the results obtained detecting the π° of 20.7 $\pm$ 0.6 percent (Roe et al. 1961; Shaklee et al. 1964). This ruled out the existence of the $\pi^{\circ\prime}$ and, by implication, the $K^{\circ\prime}$. Although this was no doubt regarded as a very improbable explanation (recall Prentki's comments earlier), the test was relatively easy to perform.

There were three further proposed explanations that cast doubt on the fundamental assumptions of modern physics and quantum mechanics. They were, therefore, in disagreement with the large amount of evidence that supported quantum mechanics, and were therefore regarded as very implausible. In fact, only one of these was tested before the consensus was reached on *CP* violation in late 1965.

It had already been noted that the observed rate for $K^{\circ}_2 \to \pi^+\pi^-$, measured at 300 K°_1 lifetimes, was that expected for K°_1 decay at 6 K°_1 lifetimes. If the exponential decay law failed for K°_1 mesons, the apparent *CP* violation could be explained as K°_1 decays (Khalfin 1965; Terent'ev 1965). This suggestion was tested, at least indirectly, by Fitch and his collaborators (Fitch, Quarles, and Wilkins 1965), who measured the K^+ lifetime in the region 1.9–7.8 mean lives and found no violation of the exponential law. This, too, was a relatively easy experiment.

The other two hypotheses, that pions were not bosons, and that the principle of superposition was violated, were not only regarded as very improbable, but also involved very difficult experimental tests. These tests were not performed until after the physics community had reached a consensus that CP was violated in late 1965. Both hypotheses were, however, eventually tested. In 1967, $K^\circ_2 \to 2\pi^\circ$, which was forbidden if pions were not bosons, was found by two groups (Gaillard et al. 1967; Cronin et al. 1967). The relevance of the observations to the test of pions as bosons was noted. The observation of K°_1–K°_2 interference without regeneration by Rubbia's group (Bohm et al. 1967) eliminated the failure of superposition alternative. This hypothesis was not even mentioned in the paper, although Kabir (1968) had discussed it in a book written before the experiments were published.

It seems clear that the strategy of combining the probability and interest of the alternative explanation with the ease of the experimental test was, in fact, used by the physics community. I might suggest, however, that the history I have recounted and the published papers themselves probably place too great an emphasis on the testing of theoretical alternatives. If such theoretical reasons are available experimenters tend to mention them. I believe that many of these experiments, including observing $K^\circ_2 \to \pi^+\pi^-$ at different energies, the search for $K^\circ_2 \to 2\pi^\circ$, and the K°_1–K°_2 interference experiments, would have been done as part of the further experimental investigation of CP violation, even if these alternatives had not existed. There is, if not a logic of experiment, a plausible sequence of measurements.

There is, however, a difference between the two episodes. In the case of parity nonconservation no alternative explanations were offered,[21] whereas in the case of CP violation numerous alternatives were considered. Why was this? The argument in the case of parity violation, in which the two classes of theories are

---

[21] This statement is strictly true only for the leptonic and semileptonic weak interactions, which involve neutrinos. There are, in fact, some nonleptonic weak decay amplitudes that do conserve parity. There were several papers that applied parity conserving schemes to the processes. For details see Franklin (1986, ch. 1, note 111).

both exclusive and exhaustive, would seem to apply just as well to the episode of *CP* violation. After all, there are only two classes of theories: those that conserve *CP* and those that do not. What is the difference between the two episodes? I believe the difference lies in the length and complexity of the derivation linking the hypothesis to the experimental result, or to the number of auxiliary hypotheses required for the derivation. In the case of parity, the experiment could be seen by inspection to violate mirror symmetry (see Figure 7.1). In the *CP* episode, what was observed was $K^{\circ}_2 \to \pi^+ \pi^-$. In order to relate this observation to the question of *CP* violation, one had to assume (1) the principle of superposition, (2) that the exponential decay law held to 300 lifetimes, (3) that the decay particles were both "real" pions and that pions were bosons, (4) that no other similar particle was produced, and (5) that there were no external conditions present that might regenerate K°_1 mesons. This is a far cry from the direct derivation in the parity experiment, and it was these auxiliary assumptions that were tested and eliminated by subsequent experiments.

Let us consider another episode from the history of physics. During the 1930s, Mott (1929, 1932), on the basis of Dirac's (1928) theory of the electron, calculated that in double scattering of electrons from heavy nuclei (high *Z*, the charge on the nucleus) there would be a forward–backward asymmetry (Franklin 1986, ch. 2). The experiment was performed many times and the conclusion reached that the expected asymmetry had not been observed. Despite this, belief in Dirac's theory remained quite strong. Mott's derivation required several assumptions including single scattering, at high velocity, off high-*Z* nuclei, at large angles and the assumption that other effects such as screening of the nucleus by atomic electrons, multiple scattering, and modifications of Coulomb's Law were negligible. It was the question of whether or not the experimental apparatuses satisfied the conditions for Mott scattering and whether or not these other effects might be present that was raised. Dirac's theory went virtually unquestioned. Why was this? I suggest that it was because of the success of the theory in predicting the existence of the positron, a result totally unexpected on the basis of background knowledge or any other existing theory, and its success in atomic spectroscopy. In comparison with these successes the discrepancy in electron scattering did not carry sufficient weight.[22] As is well known, it is very difficult to refute a well-confirmed theory. The positron provided such strong confirmation. A Bayesian analysis of this episode is given in the appendix.

Interestingly, the outcome of the story was that the experimental results were incorrect. A totally unexpected source of background electrons, due to plural scattering, precluded observation of the predicted effect.[23] This was discovered experimentally during the early 1940s and the discrepancy between theory and experiment resolved. The experimental results on double scattering of electrons were both fallible and corrigible.

This episode differs from the two discussed earlier. In the case of parity nonconservation the original experimental results were statistically overwhelming[24] and led to the overthrow of the symmetry principle. They are still regarded as correct. As I argued earlier, this is an actual solution to the weak form of the Duhem-Quine problem. In the episode of *CP* violation the original experimental evidence was strong and convincing, to at least 75 percent of the theorists who published papers on the subject. Further experiments both supported *CP* violation and eliminated, at least, by the end of 1967, all the plausible alternative explanations. This was a pragmatic solution to the Duhem-Quine problem.

In the case of the double scattering of electrons, the approach to the Duhem-Quine problem advocated here suggests that physicists were correct to question the background knowledge and the auxiliary hypotheses first. Despite numerous repetitions of the experiment and much theoretical work the discrepancy between Dirac theory and the experimental results persisted.[25] The

---

[22] There was another anomaly for Dirac theory at this time, involving the spectrum of hydrogen. This did not become important until the 1940s. For details see M. Morrison (1986).

[23] Experimenters had designed the apparatus so that the electron beams scattered from the front surface of the targets. This minimized multiple scattering. Unfortunately it also gave rise to plural scattering, which is a large-angle scatter made up of a few smaller scatterings. For more details see Franklin (1986, ch. 2).

[24] The three experiments had results that were four, thirteen, and twenty-two standard deviations from the predictions of parity conservation, respectively.

[25] This episode argues against those who believe that experimental results and

anomaly was considered to be so important that both experimental and theoretical work continued until it was resolved. No one, however, seems to have questioned the experimental results until the background effect was discovered experimentally.

This raises the interesting question of when we might reasonably question even seemingly well-confirmed experimental results. This was the case when the *V-A* theory was suggested. I have no general answer to this question. The fallibility of experimental results argues that it should always remain an option, but how probable it may be depends on the circumstances. When the *V-A* theory was first suggested the experimental evidence was contradictory, if there was a single theory that applied to all weak interactions. At the Rehovoth Conference in 1957, Konopinski had summarized the situation and noted that the angular correlation experiments on He^6, n, and Ne^{19} were consistent with the *ST* interaction, whereas those on Ne^{23}, n, Ne^{19}, and A^{35} argued for *VA*. "It seems too early to choose between them" (Konopinski, 1958, p. 330). Perhaps most importantly, the discovery of parity nonconservation and the results on the nonzero value of the parameter in muon decay were consistent only with *VA*. There were also strong theoretical reasons in favor of *VA*. If there was to be a single theory of the weak interactions it had to be *VA*. As Feynman and Gell-Mann stated, "It is amusing that this interaction [*V-A*] satisfies simultaneously almost all the principles that have been proposed on simple theoretical grounds to limit the possible β couplings. It is universal, it is symmetric, it produces two-component neutrinos [and thus explains parity violation], it conserves leptons, it preserves invariance under *CP* and *T* . . . " (Feynman and Gell-Mann 1958, pp. 197–8). This was a case in which the interest and explanatory value of the theory, in addition to its support by some of the experimental evidence, led physicists to question the experimental results.

Depending on circumstances scientists may judge it most probable and most fruitful to place the "not" with either the theory, the background knowledge, or the experimental results. I suggest that scientists examine the alternatives in the light of all existing

theory are so plastic that they can always be brought into agreement with each other. See also Galison (1987, ch. 1).

evidence, which may include the support for the theory and auxiliary hypotheses, as well as the difficulty of the experiment and the reliability of the experimental results.[26]

There are also cases in which experiment and theory may remain in conflict for a considerable time, with no resolution of the problem. Trevor Pinch (1986) has discussed the "solar-neutrino problem," which remains unsolved today, even after twenty years.

This topic has a special salience in regard to the solar-neutrino detection because there is a consensus amongst the relevant communities both as to the correctness of the theoretical prediction and as to the experimental results. There is, however, no consensus as to the consequences of the anomalous results. In the period since the first results were obtained in 1968, and despite exhaustive examination of all the relevant theoretical and experimental assumptions, no agreement has emerged as to what has gone wrong. Certainly any simple notions of the processes of how consensus is reached in science, such as the view that theories can always be modified so as to accommodate recalcitrant results or the view that troublesome experimental results can always be disregarded in the light of theory, are not adequate for dealing with this episode. In this case neither the theory nor the experiment has given way (Pinch 1986, p. 1).

I don't believe that this episode is a problem for the evidence model of science that I advocate. The difficulty seems to be with the length of time it takes to resolve an issue. Although there are some who might advocate instant rationality in science, I do not subscribe to such a view. Sometimes the solution of a problem may take a long time. This is supported by looking at the history of science. For example, the cases we have discussed have varying resolving times. In the case of parity nonconservation, it was only a period of four months between the publication of Lee and Yang's suggestion and its experimental confirmation and acceptance by the physics community. For *CP* conservation a year and a half sufficed to reach a consensus. The anomaly for Mott's calculations lasted for more than a decade. I note that it was fifty-six years from Le Verrier's 1859 announcement of an anomalous advance in the perihelion of Mercury to Einstein's general theory of relativity in 1915.

In the next chapter I will discuss an episode in which the ex-

---

[26] I am grateful to Elliott Sober for helpful discussions on this point.

perimental results themselves disagreed with one another. One result strongly supported a theory, while the other was in disagreement with the theoretical prediction. As we shall see, these results had very different evidential weights, and the disagreement was resolved after a few years.

8

Do mutants have to be slain, or do they die of natural causes? The case of atomic parity-violation experiments

In this chapter I will discuss an episode from contemporary physics, that of atomic parity violation experiments and their relation to the Weinberg-Salam unified theory of electroweak interactions. This will continue my discussion of the interaction between experiment and theory and of the epistemology of experiment. It will also explore some of the differences between my view of science and that proposed by the "strong programme" or social constructivist view in the sociology of science.[1]

This strong programme view has been summarized by Trevor Pinch.

In providing an explanation of the development of scientific knowledge, the sociologist should attempt to explain adherence to all beliefs about the natural world, whether perceived to be true or false, in a similar way (Pinch 1986, p. 3).

What is being claimed is that *many pictures* [emphasis in original] can be painted, and furthermore, that the sociologist of science cannot say that any picture is a better representation of Nature than any other (Pinch 1986, p. 8).

A central feature of this view is that change in the content of scientific knowledge is to be explained or understood in terms of the social and/or cognitive interests of the scientists involved.

There is a sense in which I am in agreement with this symmetrical view. The evidence model I have suggested explains adherence to scientific beliefs in terms of their relationship to valid experimental evidence. There is an asymmetry in my view

---

[1] I already began this discussion in Chapter 6 when I considered the differences between Andy Pickering and myself on the question of Morpurgo's experiments.

because "true" beliefs[2] agree with the experimental evidence whereas false beliefs do not, but the underlying methodology, that of comparison to experimental evidence, is the same. Obviously I do not agree with the social constructivists that all pictures of the world are equally good. In 1958, the *V-A* theory of weak interactions was clearly better, or in my terms more in agreement with the experimental evidence, than any of its competitors.

I suspect that the social constructivists would, however, argue that I have considered the wrong evidence. They would say that it is not the experimental results, but rather the social and/or cognitive interests of the scientists, that must be used in the explanation. The question that the reader must answer is, which of us is telling the more plausible story?

Some theoretical difficulties have been raised about the social constructivist view. The first is that of reflexivity. If the social constructivists argue that "All beliefs are relative" then this must apply to their own view. Why then should we believe them? Pinch (1986, ch. 1) offers several responses to this argument, which I find unpersuasive, but rightly insists that deciding empirical questions about science by logic is, at the very least, suspicious. I note that for methodological questions Pinch and other social constructivists seem to subscribe to an evidence model.

Woolgar (1981) has argued that the social constructivists assume that "interests" are given unproblematically, rather than being themselves socially constructed. He has also raised questions concerning the way in which these interests are used to explain the content of science. He argues that the causal relationship between the interests and the scientific beliefs is asserted rather than argued for.

I do not find these theoretical arguments convincing. I believe that the evidence model applies to both science and to the study of science.[3] The question is whether the evidence model or the social constructivist view accords better with the actual history

---

[2] I would prefer to speak about reasonably warranted beliefs rather than true beliefs, but this is the term used by Pinch.

[3] Some readers may worry that I am using the evidence model to decide whether or not an evidence model applies to science. I don't think this is a serious problem. Using the evidence model in no way guarantees that the view that scientists use such a model will be supported by the evidence.

and practice of science. Unfortunately, studies of the same episode from these two different points of view are rare. The only
previous case[4] I know of is that of the discovery of weak-neutral
currents, studied by both Galison (1987) and Pickering (1984b).[5]

During the 1960s events now attributed to weak-neutral currents were seen but were attributed to neutron background. At
the time there was no theoretical prediction of such currents.
Later, after theory had predicted the existence of these currents,
new experiments reported their existence and the earlier experiments were reinterpreted. Pickering explains this as the mutual
adjustment of the theory of the phenomena, the experimental
apparatus, and the theory of that apparatus. Galison's study shows
that enormous effort went into producing "arguments that would
stand up in court" for the existence of neutral currents. This is
not to say that the theory of the phenomena was not a guide for
the experimenters, but only that theory cannot produce events in
bubble or spark chambers. I find Galison's discussion, which I
believe supports the evidence model, more persuasive. I also think
that the criticism offered in chapter 6 of Pickering's discussion of
Morpurgo's experiments also applies to his discussion of the weak-
neutral current experiments. I believe that he underestimates the
importance of the arguments used to establish the validity of an
experimental result.

I now present a second comparison of these opposing views of
experiment.

In his recent book, *Constructing Quarks* (1984a), Pickering
discussed the early experiments on atomic parity violation,
which were anomalous for the Weinberg-Salam (W-S) unified
theory of electroweak interactions. These experiments, performed at Oxford University and at the University of Washington and published in 1976 and 1977, measured the parity
nonconserving optical rotation in atomic bismuth. The results
disagreed with the predictions of the Weinberg-Salam theory.

---

[4] Both Pinch (1986) and Shapere (1982) have discussed the solar-neutrino experiment, but only Pinch presents a detailed history. Shapere is primarily
concerned with the philosophical issue of observation.

[5] I will only briefly summarize the arguments here. I encourage the reader to
examine both of these studies and decide which model seems to fit the episode
better. Hones (1987) discusses this episode from a philosophical view and
concludes that the scientific community behaved rationally.

Another experiment, performed at the Stanford Linear Accelerator Center in 1978, on the scattering of polarized electrons from deuterons, confirmed the theory. Pickering regards the Oxford and Washington experiments as mutants, slain by the SLAC experiment.[6] By 1979 the Weinberg-Salam theory was regarded by the high-energy physics community as established, despite the fact that as Pickering recounts, "there had been no *intrinsic* change [emphasis in original] in the status of the Washington-Oxford experiments" (Pickering 1984a, p. 301). In Pickering's view, "particle physicists *chose* [emphasis in original] to accept the results of the SLAC experiment, *chose* to interpret them in terms of the standard model (rather than some alternative which might reconcile them with the atomic physics results), and therefore *chose* to regard the Washington-Oxford experiments as somehow defective in performance or interpretation" (Pickering 1984a, p. 301). The implication seems to be that these choices were made so that the experimental evidence would be consistent with the Standard Model,[7] and that there weren't good, independent reasons for them.[8]

Pickering regards this episode as supporting his view that "there is no obligation upon anyone framing a view of the world to take account of what twentieth century science has to say" (Pickering 1984a, p. 413). He obviously doubts that science is a reasonable enterprise based on valid experimental or observational evidence. In this chapter I will reexamine the history of this episode, presenting both Pickering's interpretation and an alternative explanation of my own, arguing that there were good reasons for the decision of the physics community. I will also look at the subsequent history of these measurements, along with other atomic physics experiments that have relevance for the Weinberg-Salam theory.

---

[6] The detailed history, which follows, is far more complex. Other experimental groups in Novosibirsk and Berkeley found atomic physics results in agreement with the W-S theory, while a Moscow group confirmed the anomaly.

[7] The Standard Model includes quarks and the Weinberg-Salam theory.

[8] To be fair, Pickering, as will be discussed subsequently, couches his discussion in terms of the adjustment of theoretical and experimental research practice. I believe, however, that in this case his meaning is essentially agreement between experiment and theory. In any event, as discussed below, I believe that he made an incorrect judgment on the relative evidential weight of the two different experiments.

8.1 THE EARLY EXPERIMENTS

In 1957, it had been experimentally demonstrated that parity, or left–right symmetry, was violated in the weak interactions.[9] This feature of the weak interactions had been incorporated into the Weinberg-Salam theory. The theory predicted that one would see weak-neutral-current effects in the interactions of electrons with hadrons, strongly interacting particles. The effect would be quite small when compared to the dominant electromagnetic interaction, but could be distinguished from it by the fact that it violated parity conservation.[10] A demonstration of such a parity-violating effect and a measurement of its magnitude would test the W-S theory.[11]

One such predicted effect was the rotation of the plane of polarization of polarized light when it passed through bismuth vapor. Such a rotation is possible only if parity is violated. This was the experiment performed by the Oxford and Washington groups. They both used bismuth vapor but used light corresponding to different transitions in bismuth, λ = 648 nm (Oxford) and λ = 876 nm (Washington). They published a joint preliminary report noting, "we feel that there is sufficient interest to justify an interim report" (Baird et al. 1976, p. 528). They reported values for R, the parity violating parameter, of $R = (-8 \pm 3) \times 10^{-8}$ (Washington) and $R = (+10 \pm 8) \times 10^{-8}$ (Oxford). "We conclude from the two experiments that the optical rotation, if it exists, is smaller than the values -3×10^{-7} and -4×10^{-7} predicted by the Weinberg-Salam model plus the atomic central field approximation" (Baird et al. 1976, p. 529).

Pickering offers the following interpretation.

---

[9] See Chapter 4 for details.

[10] M. Bouchiat and C. Bouchiat (1974).

[11] If parity is violated, but time-reversal invariance holds, as experiment suggests, then there will be an electric dipole transition element between two states of the same parity, such as two $S_{1/2}$ states of cesium. All experiments thus far have been based on measurements of an electroweak interference between such a weak interaction amplitude, E_1^{PNC}, which does not conserve parity, and a parity conserving electromagnetic amplitude. In the optical-rotation experiments on bismuth and lead, one observes an interference between the allowed M_1 (magnetic dipole) amplitude and the E_1^{PNC} amplitude. In highly forbidden M_1 transitions in heavy elements such as thallium or cesium, one observes an interference between the parity-nonconserving amplitude and the Stark-induced (by an electric field) amplitude. For details see Bouchiat and Pottier (1984).

The *caveat* to this conclusion was important. Bismuth had been chosen for the experiment because relatively large effects were expected for heavy atoms, but when the effect failed to materialise a drawback of the choice became apparent. To go from the calculation of the primitive neutral-current interaction of electrons with nucleons to predictions of optical rotation in a real atomic system it was necessary to know the electron wave-functions, and in a multielectron atom like bismuth these could only be calculated approximately. Furthermore, these were novel experiments and it was hard to say in advance how adequate such approximations would be for the desired purpose. Thus in interpreting their results as a contradiction of the Weinberg-Salam model the experimenters were going out on a limb of *atomic* theory. Against this they noted that four independent calculations of the electron wave functions had been made, and that the results of these calculations agreed with one another to within twenty-five percent. This degree of agreement the experimenters found "very encouraging" although they conceded that "Lack of experience of this type of calculation means that more theoretical work is required before we can say whether or not the neglected many-body effects in the atomic calculation would make R [the parity-violating parameter] consistent with the present experimental limits" (Pickering 1984a, pp. 295–6).

Pickering attributes all of the uncertainty in the comparison between experiment and theory to the theoretical calculations and none to the experimental results themselves.

The comparison was even more uncertain than Pickering implies and included the uncertainty in the experimental results. The experimenters reported that the "quoted statistical error represents 2 s.d. [standard deviations]. There are, however, also systematic effects which we believe do not exceed $\pm 10 \times 10^{-8}$, but which are not yet fully understood" (Baird et al. 1976, p. 529). Thus, there were possible systematic experimental uncertainties of the same order of magnitude as the expected effect. As Pickering states, these were novel experiments, using new and previously untried techniques. This also tended to make the experimental results uncertain.[12]

The theoretical calculations of the expected effect were also uncertain. The Oxford-Washington joint paper noted that Khriplovich had argued, in a soon to be published paper, that the approximate theory overestimated R by a factor of approximately

---

[12] The Washington group had, in fact, published an earlier paper in which they merely set an upper limit for the parity-violating amplitude of 10^{-6} (Soreide et al. 1976).

Table 8.1.

Method	R (10^{-7})	Reference
Hartree-Fock	-2.3	Brimicombe, Loving, and Sandars (1976)
Hartree-Fock	-3.5	Henley and Wilets (1976)
Semiempirical	-1.7	Novikov, Sushkov, and Khriplovich (1976)
Multiconfiguration	-2.4	Grant (1976, private communication)

1.5. In addition the four calculations agreed with their mean only to within approximately $\pm 25\%$. This made the largest and smallest calculated values of R differ by almost a factor of 2. The Washington group reported, in their subsequent 1977 paper, that the calculated values of R were those given in Table 8.1. By this time, the quoted uncertainty in the calculations had increased to $\pm 30\%$.

In September, 1977, both the Washington and Oxford groups published more detailed accounts of their experiments with somewhat revised results (Lewis et al. 1977; Baird et al. 1977). Both groups reported results in substantial disagreement with the predictions of the Weinberg-Salam theory, although the Washington group stated, "more complete calculations that include many-particle effects are clearly desirable" (Lewis et al. 1977, p. 795). The Washington group reported a value of $R = (-0.7 \pm 3.2) \times 10^{-8}$, which was in disagreement with the prediction of approximately -2.5×10^{-7} (see Table 8.1). This value is inconsistent with their earlier result of $(-8 \pm 3) \times 10^{-8}$, where the stated uncertainty was 2 standard deviations. The new uncertainty was 1.5 s.d. The difference between the two values is $(7.3 \pm 2.5) \times 10^{-8}$, a 2.9 s.d.-effect, which has a probability of 0.37% of being equal to 0, an unlikely occurrence. This inconsistency was not discussed by the experimenters, but it was discussed within the atomic physics community and lessened the credibility of the result.[13] The uncertainty of the result (the standard deviation) had also increased from 1.5×10^{-8} to 2.1×10^{-8}. The Oxford result was $R = (+2.7 \pm 4.7) \times 10^{-8}$, again in disagreement with the Weinberg-Salam prediction of approximately -30×10^{-8}. They

---

[13] Carl Wieman, who made important contributions to the later atomic physics experiments, recalls such discussions (private communication).

noted, however, that there was a systematic effect in their apparatus. They found a change in ϕ_r, the rotation angle, due to slight misalignment of the polarizers, optical rotation in the windows, and so forth, of order 2×10^{-7} radians. "Unfortunately, it varies with time over a period of minutes, and depends sensitively on the setting of the laser and the optical path through the polarizer. While we believe we understand this effect in terms of imperfections in the polarizers combined with changes in laser beam intensity distribution, we have been unable to reduce it significantly" (Baird et al. 1977, p. 800). A systematic effect of this size certainly cast doubt on the result.

Pickering reported that the papers also "described two 'hybrid' unified electroweak models, which used neutral heavy leptons to accommodate the divergence with the findings of high energy neutrino scattering" (Pickering 1984a, p. 297). Although such models were discussed in the literature at the time (see the discussion to follow), there is no mention of such speculation in these two experimental papers.

How were these results viewed by the physics community? In the same issue of *Nature,* in which the joint paper was published, Frank Close, a particle theorist, summarized the situation. "Is parity violated in atomic physics? According to experiments being performed independently at Oxford and the University of Washington the answer may well be no. . . . This is a very interesting result in light of last month's report . . . claiming that parity is violated in high energy 'neutral-current' interactions between neutrinos and matter" (Close 1976, p. 505).

The experiment that Close referred to had concluded, "Measurements of R^v and $R^{\bar{v}}$, the ratios of neutral current to charged current rates for v and $\bar{v}$ cross sections, yield neutral current rates for v and $\bar{v}$ that are consistent with a pure V-A interaction but 3 standard deviations from pure V or pure A, indicating the presence of parity nonconservation in the weak neutral current" (Benvenuti et al. 1976, p. 1039).

Close noted that as the atomic physics results stood, they appeared to be inconsistent with the predictions of the Weinberg-Salam model supplemented by atomic physics calculations. He also remarked that, "At present the discrepancy can conceivably be the combined effect of systematic effects in atomic physics calculations and systematic uncertainties in the experiments"

(Close 1976, pp. 505–6). Pickering states, "if one accepted the Washington-Oxford result, the obvious conclusion was that neutral current effects violated parity conservation in neutrino interactions and conserved parity in electron interactions" (Pickering 1984a, p. 296). Close discussed this possibility along with another alternative that had an unexpected (on the basis of accepted theory) energy dependence, so that the high-energy experiments (the neutrino interactions) showed parity nonconservation whereas the low-energy atomic physics experiments would not. "Whether such a possibility could be incorporated into the unification ideas is not clear. It also isn't clear, yet, if we have to worry. However, the clear blue sky of summer now has a cloud in it. We wait to see if it heralds a storm" (Close 1976, p. 506).

In Pickering's view, the 1977 publication of the Oxford and the Washington results indicated that "the storm that Frank Close had glimpsed had materialised and was threatening to wash away the basic Weinberg-Salam model, although not the gauge-theory enterprise itself" (Pickering 1984a, p. 298).[14] There is some support for this in a summary of the Symposium on Lepton and Photon Interactions at High Energies, held in Hamburg August 25–31, 1977, given by David Miller in *Nature* (1977). Miller noted that Sandars had reported that neither his group at Oxford nor the Washington group had seen any parity violating effects and that "they have spent a great deal of time checking both their experimental sensitivity and the theory in order to be sure" (Miller 1977, p. 288). Miller went on to state (as Pickering also reported) that "S. Weinberg and others discussed the meaning of these results. It seems that the $SU(2) \times U(1)$ is to the weak interaction what the naive quark-parton model has been to QCD, a first approximation which has fitted a surprisingly large amount of data. Now it will be necessary to enlarge the model to accommodate the new quarks and leptons, the absence of atomic neutral currents, and perhaps also whatever it is that is causing trimuon events" (Miller 1977, p. 288). Nevertheless I believe that the uncertainty in these experimental results made the disagreement with the W-S theory only a worrisome situation and not a crisis

---

[14] Pickering also discusses two other anomalies for the Weinberg-Salam model, the high-y anomaly and the trimuon events. These have also been shown experimentally to be incorrect.

as Pickering believes.[15] In any event, the monopoly of Washington and Oxford was soon broken.

The experimental situation changed in 1978 when Barkov and Zolotorev (1978a, 1978b, and 1979), two Soviet scientists from Novosibirsk, reported measurements on the same transition in bismuth as the Oxford group. Their results agreed with the predictions of the W-S model. They gave a value for $\psi_{exp}/\psi_{W\text{-}S}$ = $(+1.4 \pm 0.3)$ k, where ψ was the angle of rotation of the plane of polarization by the bismuth vapor. "The factor k was introduced because of inexact knowledge of the bismuth vapor, and also because of some uncertainty in the relative positions of the resonator modes and the absorption-line centers. According to our estimate, the factor lies in the interval from 0.5 to 1.5" (Barkov and Zolotorev 1978a, p. 360). They concluded that their result "does not contradict the predictions of the Weinberg-Salam model." A point to be emphasized here is that agreement with theoretical prediction depended (and still does depend) on which method of calculation one chose, as discussed earlier. A somewhat later paper (Barkov and Zolotorev 1978b) changed the result to $\psi_{exp}/\psi_{W\text{-}S} = 1.1 \pm 0.3$.

Subsequent papers, in 1979 and 1980 (Barkov and Zolotorev 1979, 1980b), reported more extensive data and found a value for $R_{exp}/R_{theor} = 1.07 \pm 0.14$. They also reported that the latest unpublished results from the Washington and Oxford groups, which had been communicated to them privately, showed parity violation, although "the results of their new experiments have not reached good reproducibility" (Barkov and Zolotorev 1979, p. 312). These later results were also presented at the 1979 conference, discussed later, at which Dydak reviewed the situation.

According to Pickering, "the details of the Soviet experiment were not known to Western physicists, making a considered evaluation of its result problematic" (1984a, p. 299). This is simply not correct. During September 1979, an international workshop devoted to neutral-current interactions in atoms was held in Cargese (Williams 1980). This workshop was attended by representatives of virtually all of the groups actively working in the field, including Oxford, Washington, and Novosibirsk. At that

---

[15] My view is supported by the recollections of Carl Wieman and of James Scott, a physicist visiting at Oxford at the time of these experiments.

workshop not only did the Novosibirsk group present a very detailed account of their experiment (Barkov and Zolotorev 1980a), but, as C. Bouchiat remarked in his workshop summary paper, "Professor Barkov, in his talk, gave a very detailed account of the Novosibirsk experiment and answered many questions concerning possible systematic errors" (C. Bouchiat 1980, p. 364). In addition, an examination of both Soviet publications indicates that they contained as much technical detail as the 1977 Oxford and Washington publications, and far more detail than the joint 1976 paper. There was also communication between the Soviet and Oxford groups. The Soviets reported that they had been able to uniquely identify the hyperfine structure of the 6477Å (648 nm) line of atomic bismuth and that "the results of these measurements agree also with the results in Oxford (P. Sandars, private communication)" (Barkov and Zolotorev 1978a, p. 359).

In early 1979, a Berkeley group reported an atomic physics result for thallium that agreed with the predictions of the W-S model (Conti et al. 1979). They investigated the polarization of light passing through thallium vapor and found a circular dichroism $\delta = (+5.2 \pm 2.4) \times 10^{-3}$ in comparison with the theoretical prediction of $(+2.3 \pm 0.9) \times 10^{-3}$. Although these were not definitive results – they were only two s.d. from zero – they did agree with the model in both sign and magnitude.

It seems fair to say that in mid–1979 the atomic physics results concerning the Weinberg-Salam theory were inconclusive. The Oxford and Washington groups had originally reported a discrepancy, but their more recent results, although preliminary, showed the presence of the predicted parity nonconserving effects. The Soviet and Berkeley results agreed with the model. Dydak (1979) summarized the situation in a talk at a 1979 conference. "It is difficult to choose between the conflicting results in order to determine the *eq* [electron–quark] coupling constants. Tentatively, we go along with the positive results from Novosibirsk and Berkeley groups and hope that future development will justify this step (it cannot be justified at present, on clear-cut experimental grounds)" (1979, p. 35).

Pickering states, "Having decided not to take into account the Washington-Oxford results, Dydak concluded that parity violation in atomic physics was as predicted in the standard model" (1984a, p. 300).

I find little justification for Pickering's conclusion. Dydak was, as Pickering himself notes, attempting to determine the best values for the parameters describing neutral-current electron scattering. Dydak had tentatively adopted the results in agreement with the W-S model, admitting that experiment did not, at the time, justify this. He concluded nothing about the validity of the Standard Model. Bouchiat, in his summary paper discussed earlier, was more positive. After reviewing the Novosibirsk experiment as well as the conflict between the earlier and later Washington and Oxford results he remarked, *"As a conclusion on this Bismuth session, one can say that parity violation has been observed roughly with the magnitude predicted by the Weinberg-Salam theory"* (C. Bouchiat 1980, p. 365, emphasis in original).[16] Even this more positive statement does not conclude that the results agree with the predictions of the theory, but states only that the experimental results were of the correct order of magnitude.

The situation was made even more complex when a group at the Stanford Linear Accelerator Center (SLAC) reported a result on the scattering of polarized electrons from deuterium that agreed with the W-S model (Prescott et al. 1978, 1979). This was the E122 experiment, discussed by Pickering. They not only found the predicted scattering asymmetry but also obtained a value for $\sin^2\theta_W = 0.20 \pm 0.03$ (1978) and 0.224 ± 0.020 (1979) in agreement with other measurements. ($\sin^2\theta_W$ is an important parameter in the W-S theory.) "We conclude that within experimental error our results are consistent with the W-S model, and furthermore our best value of $\sin^2\theta_W$ is in good agreement with the weighted average for the parameter obtained from neutrino experiments" (Prescott et al. 1979, p. 528).

Pickering concluded his story as follows.

In retrospect, it is easy to gloss the triumph of the standard model in the idiom of the 'scientist's account': the Weinberg-Salam model, with an appropriate complement of quarks and leptons, made predictions which were verified by the facts. But missing from this gloss, as usual, is the element of choice. In assenting to the validity of the standard

---

[16] For another summary of the experimental situation at the time see Commins and Bucksbaum (1980). They regarded the situation with respect to the bismuth results as unresolved.

model, particle physicists chose to accept certain experimental reports and to reject others. The element of choice was most conspicuous in the communal change of heart over the Washington-Oxford atomic physics experiments and I will focus on that episode here. We saw in the preceding section that in 1977 many physicists were prepared to accept the null results of the Washington and Oxford experiments and to construct new electroweak models to explain them. We also saw that by 1979 attitudes had hardened. In the wake of experiment E122, the Washington-Oxford results had come to be regarded as unreliable. In analysing this sequence, it is important to recognise that between 1977 and 1979 there had been no *intrinsic* change in the status of the Washington-Oxford experiments. No data were withdrawn, and no fatal flaws in the experimental practice of either group had been proposed. What had changed was the *context* within which the data were assessed. Crucial to this change of context were the results of experiment E122 at SLAC. In its own way E122 was just as innovatory as the Washington-Oxford experiments and its findings were, in principle, just as open to challenge. But particle physicists *chose* to accept the results of the SLAC experiment, *chose* to interpret them in terms of the standard model (rather than some alternative which might reconcile them with the atomic physics results) and therefore *chose* to regard the Washington-Oxford experiments as somehow defective in performance or interpretation (Pickering 1984a, p. 301).

Though I do not dispute Pickering's contention that choice was involved in the decision to accept the Weinberg-Salam model, I disagree with him about the reasons for that choice. In Pickering's view, "The standard electroweak model unified not only the weak and electromagnetic interactions: it served also to unify practice within otherwise diverse traditions of HEP [high-energy physics] theory and experiment. . . . Matched against the mighty traditions of HEP, the handful of atomic physicists at Washington and Oxford stood little chance" (Pickering 1984a, pp. 301–2). In my view, the choice was a reasonable one based on convincing, if not overwhelming, experimental evidence. As we shall see, this was even agreed to by atomic physicists.

The issue seems to turn on the relative evidential weight one assigns to the original Oxford and Washington atomic physics results and to the SLAC E122 experiment on the scattering of polarized electrons. Pickering seems to regard them as having equal weight. I do not. I argued earlier that both the original experimental results on bismuth, as well as the comparison between experiment and theory, were quite uncertain. When one adds to this original uncertainty the later contradictory results of both the Washington and Oxford groups, the parity nonconserving measurement of the Novosibirsk group, and the Berkeley result on thallium, the original results were, at the very least, very uncertain.[17]

In 1981, the Washington group published their latest result "that agrees in sign and approximate magnitude with recent calculations based upon the Weinberg-Salam theory" (Hollister et al. 1981, p. 643). Although theoretical calculations had reduced the size of the expected effect by a factor of approximately 3 or 4 (depending on which method of calculation one chose) the theoretical results still did not agree with the 1977 Washington-Oxford measurements. In Pickering's words, "Instead, the 1981 Washington recantation was based upon new data taken with new apparatus, and it remained the case that no defect of interpretation or performance of their 1977 experiment had been identified. Significantly, though, the Washington group sought to legitimate their new positive findings by casting doubt upon the earlier reports from atomic physics experiments, including their own" (Pickering 1984a, p. 307).

I discussed earlier the uncertainty introduced into the 1977 Washington and Oxford results by systematic effects. This is made even more evident when we examine the 1980 report by a group from Moscow. They reported measurements on the same transition in bismuth (λ = 648 nm) that the Oxford group had used (Bogdanov 1980a, 1980b).[18] Their measurement was in disagreement with the predictions of the W-S theory. They reported an optical rotation due to the parity nonconserving interaction of $\Delta\phi_{PNC} = (-0.22 \pm 1.0) \times 10^{-8}$ rad, in disagreement with the theoretical prediction of 10^{-7} rad. They discussed two sources of systematic errors that could give rise to effects of the same size as those expected from parity nonconservation: variation in laser intensity due to scanning the laser frequency and interference between the main laser beam and scattered light. They also dis-

---

[17] See Commins and Bucksbaum (1980) and C. Bouchiat (1980) for summaries of the situation by atomic physicists, not members of the high-energy physics community.

[18] This result was also presented as an addendum to the Proceedings of the Cargese Workshop (Williams 1980).

Table 8.2.

Experimental series	$\Delta\phi_{PNC}$ (in 10^{-8} rad)
1	(-1.52 ± 2.1)
2	$(+5.41 \pm 3.5)$
3	(-0.16 ± 2.4)
4	(-1.96 ± 2)
5	(-6.76 ± 2.6)
6	$(+3.70 \pm 2.4)$

cussed the measures taken to reduce the errors due to these effects. Even so, they reported the results for their six measurement series that appear in Table 8.2.

They remarked that the spread in these individual series substantially exceeded the error in their quoted result and attributed that to time-dependent instrumental errors. Once again, there were systematic errors in this type of experiment that were approximately the same size as the effects predicted by the W-S theory.

The 1981 Washington paper stated, "Our experiment and the bismuth optical-rotation experiments by three other groups [Oxford, Moscow, and Novosibirsk] *have yielded results with significant mutual discrepancies far larger than the quoted errors* (Hollister et al. 1981, p. 643, emphasis added). They also pointed out, as I discussed earlier, that their earlier measurements "were not mutually consistent."

The moral of the story is clear. These were extremely difficult experiments, beset with systematic errors of approximately the same size as the predicted effects. There is no reason to give priority to the earliest measurements, as Pickering does. One might suggest, rather, that these earlier results were perhaps less reliable because not all of the systematic errors were known.

I will now examine the arguments presented by the SLAC group in favor of the validity and reliability of their measurement. I agree with Pickering that "in its own way E122 was just as innovatory as the Washington-Oxford experiments and its findings were, in principle, just as open to challenge" (Pickering 1984a, p. 301). For this reason, the SLAC group presented a very detailed discussion of their experimental apparatus and result and performed many checks on their experiment.

The experiment depended, in large part, on a new high-intensity source of longitudinally polarized electrons. The polarization of the electron beam could be varied by changing the voltage on a Pockels cell. "This reversal was done randomly on a pulse-to-pulse basis. The rapid reversals minimized the effects of drifts in the experiment, and the randomization avoided changing the helicity synchronously with periodic changes in experimental parameters" (Prescott et al. 1978, p. 348). It had been demonstrated, in an earlier experiment, that polarized electrons could be accelerated with negligible depolarization. In addition, both the sign and magnitude of the beam polarization were measured periodically by observing the known asymmetry in elastic electron–electron scattering from a magnetized iron foil.

The experimenters also checked whether or not the apparatus produced spurious asymmetries. They measured the scattering using the unpolarized beam from the regular SLAC electron gun, for which the asymmetry should be zero. They assigned polarizations to the beam using the same random number generator that determined the sign of the voltage on the Pockels cell. They obtained a value for $A_{exp}/P_e = (-2.5 \pm 2.2) \times 10^{-5}$, where A_{exp} was the experimental asymmetry and P_e was the beam polarization for the polarized source, $P_e = 0.37$. This was consistent with zero and demonstrated that the apparatus could measure asymmetries of the order of 10^{-5}.

They also varied the polarization of the beam by changing the angle of a calcite prism thereby changing the polarization of the light striking the Pockels cell. They expected that $A_{exp} = |P_e| A \cos(2\phi_p)$, where ϕ_p was the prism angle. The results are shown in Figure 8.1. Not only do the data fit the expected curve, but the fact that the results at 45° are consistent with zero indicates that other sources of error in A_{exp} are small. The graph shows the results for two different detectors, a nitrogen-filled Cerenkov counter and a lead glass shower counter. The consistency of the results increases the belief in the validity of the measurements. "Although these two separate counters are not statistically independent, they were analyzed with independent electronics and respond quite differently to potential backgrounds. The consis-

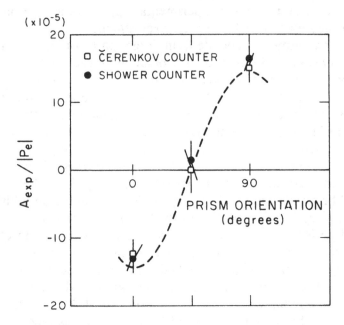

Figure 8.1. Experimental asymmetry as a function of prism angle for both the Cerenkov counter and the shower counter. The dashed line is the predicted behavior. From Prescott et al. (1978).

tency between these counters serves as a check that such backgrounds are small" (Prescott et al. 1978, p. 350).

The electron beam helicity also depended on E_o, the beam energy, because of the g-2 precession of the spin as the electrons passed through the beam transport magnets. The expected distribution as well as the experimental data for $A_{exp}/|P_e|Q^2$ is shown in Figure 8.2. Q^2 is the square of the momentum transfer. "The data quite clearly follow the g-2 modulation of the helicity," and the fact that the value at 17.8 GeV is close to zero demonstrates that any transverse spin effects were small.

A serious source of potential error came from small systematic differences in the beam parameters for the two helicities. Small changes in beam position, angle, current, or energy could influence the measured yield, and if correlated with reversals of beam helicity could cause apparent, but spurious, parity-violating asymmetries. These quantities were carefully monitored and a feedback system was used to stabilize them. "Using the measured pulse-to-pulse beam information together with the measured sen-

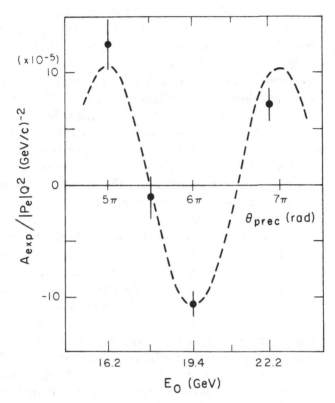

Figure 8.2. Experimental asymmetry as a function of beam energy. The expected behavior is the dashed line. From Prescott et al. (1978).

sitivities of the yield to each of the beam parameters, we made corrections to the asymmetries for helicity dependent differences in beam parameters. For these corrections, we have assigned a systematic error equal to the correction itself. The most significant imbalance was less than one part per million in E_0 [the beam energy] which contributed -0.26×10^{-5} to A/Q^2" (Prescott et al. 1978, p. 351). This is to be compared to their final result of $A/Q^2 = (-9.5 \pm 1.6) \times 10^{-5}$ $(\text{GeV}/c)^{-2}$. This was regarded by the physics community as a reliable and convincing result.[19]

---

[19] The experimenters used several strategies to establish the validity of their result that I have discussed earlier as parts of an epistemology of experiment. The experimenters intervened and observed the predicted effects when they changed the angle of the calcite prism and when they varied the beam energy. They checked and calibrated their apparatus by using the unpolarized SLAC beam and observed no instrumental asymmetries and found that their ap-

Contrary to Pickering's claim, hybrid models were both considered and tested by E122. In their first paper they pointed out that the hybrid model was consistent with their data only for values of $\sin^2\theta_W < 0.1$, which was inconsistent with the measured value of approximately 0.23. In the second paper (Prescott et al. 1979) they plotted their data as a function of $y = (E_o - E')/E_o$, where E' is the energy of the scattered electron. Both models, W-S and the hybrid, made definite predictions for this graph. The results are shown in Figure 8.3 and the superiority of the W-S model is obvious. For W-S they obtained a value of $\sin^2\theta_W = 0.224 \pm 0.020$ with a χ^2 probability of 40%. The hybrid model gave a value of 0.015 and a χ^2 probability of 6×10^{-4}, "which appears to rule out this model."

My interpretation of this episode differs drastically from Pickering's. The physics community chose to accept an extremely carefully done and carefully checked experimental result that confirmed the Weinberg-Salam theory. This view is supported by Bouchiat's 1979 summary. After hearing a detailed account of the SLAC experiment by Prescott, he stated, "To our opinion, this experiment gave the first truly convincing evidence for parity violation in neutral current processes" (C. Bouchiat 1980, p. 358). "I would like to say that I have been very much impressed by the care with which systematic errors have been treated in the experiment. It is certainly an example to be followed by all people working in this very difficult field" (C. Bouchiat 1980, pp. 359–60).[20] The physics community chose to await further developments in the atomic parity-violating experiments, which, as I have shown, were uncertain. Pickering is correct that in stating that there had been no intrinsic change in the early Washington and Oxford results. They began as uncertain, although worrisome, and remained uncertain. The subsequent history (discussed in the next section) shows that although other reliable atomic physics experiments confirm the W-S theory, the bismuth results, al-

paratus could measure asymmetries of the expected size. They also used different counters, the lead glass shower counter and the gas Cerenkov counter, and obtained independent confirmation of the validity of their measurement.

[20] Bouchiat was a theoretical atomic physicist working on parity violation, and not a member of the high-energy physics community. As one can see, it was not the mighty traditions of high-energy physics that convinced him, but rather the experimental evidence.

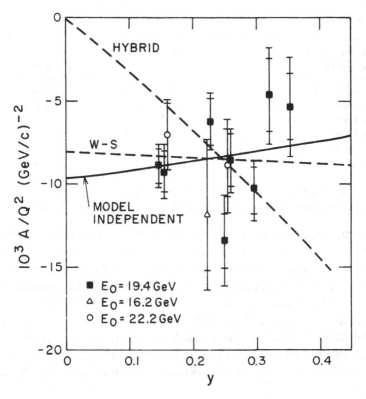

Figure 8.3. Asymmetries measured at three different energies plotted as a function of $y = (E_0 - E')/E_0$. The predictions of the hybrid model, the Weinberg-Salam theory, and a model independent calculation are shown. "The Weinberg-Salam model is an acceptable fit to the data; the hybrid model appears to be ruled out" (Prescott et al. 1979).

though generally in agreement with the predictions, are still somewhat uncertain. In addition, the most plausible alternative to the W-S model, that could reconcile the original atomic physics results with the electron scattering data, was tested and found wanting. There certainly was a choice made, but, as the "scientist's account" or evidence model suggests, it was made on the basis of experimental evidence. The mutants died of natural causes.

8.2 THE LATER EXPERIMENTS

In this section I will discuss the subsequent history of the bismuth experiments, along with a brief treatment of the other atomic-

Table 8.3.

Bismuth optical activity (λ = 648 nm)	Experiment ($R \times 10^8$)	Theory	
Novosibirsk	-20.6 ± 3.2	Novosibirsk	-19
Oxford I	-10.3 ± 1.8	Oxford	-14
Oxford II	-11.2 ± 4.1		
Old Oxford	2.7 ± 4.7		
(λ = 876 nm)			
Washington	$-10 \pm 2 \pm$?	Novosibirsk	-14
		Oxford	-12
Old Washington	-0.7 ± 3.2		

physics parity-violation experiments that have relevance for the W-S theory. If, as Pickering claims, one can choose the experimental result one prefers, then there was no need for this subsequent work because the decision had already been made. If, as the evidence model suggests, one is interested in valid and consistent measurements so that one can establish whether or not the experimental results confirm or refute the accepted theory, test the theory in different phenomena and at different energies, and eliminate competing hypotheses, then all this activity makes sense.

I begin with C. Bouchiat's 1979 summary of the situation (C. Bouchiat 1980). He presented Table 8.3. He remarked that the Novosibirsk result had been published, whereas both the Washington and Oxford results were in the nature of progress reports on recent trends in their experiments, not definite results. He also noted that there was no explanation of the large difference between the old and new Washington and Oxford results, and that there was a factor of two discrepancy between the Novosibirsk and Oxford results at λ = 648 nm. The difference in both theoretical approach and the numerical value of the calculation between the two groups was also mentioned.

In 1981, the Washington group published their most recent measurement on the optical rotation in bismuth at λ = 876 nm (Hollister et al. 1981). The theoretical calculations they used to compare with their data ranged from $R = -8 \times 10^{-8}$ to -17×10^{-8}. Their value of $R = (-10.4 \pm 1.7) \times 10^{-8}$ agreed in "sign and approximate magnitude with recent calculations of the effect in bismuth based on the Weinberg-Salam theory." They pointed out that since their earliest measurements they had "added a new laser, improved the optics, and included far more extensive experimental checks" (Hollister et al. 1981, p. 643). They excluded these first three measurements from their average because they were made without the new systematic checks and controls, which were to be discussed in detail in a forthcoming paper.

This discussion appeared in an extensive review of atomic parity-violation experiments by Fortson and Lewis (1984), two members of the Washington group. They reported experimental controls on both the polarizer angle and the laser frequency. They also used alternate cycles, in which their bismuth oven was turned off, to avoid a spurious effect that could mimic the expected parity-nonconserving effect. They also examined their data for any correlations between the measured values of the parity-nonconserving parameter and any other experimental variables. This procedure set limits on known sources of systematic error and "initially helped to uncover some errors and eliminate them." Included in these errors were those due to wavelength dependent effects and to beam movement. As we shall see, these were problems in all of these experiments.

The Novosibirsk group's results did not change very much from the value cited above by C. Bouchiat. Their last published measurement (Barkov and Zolotorev 1980b) gave $R = (-20.2 \pm 2.7) \times 10^{-8}$, which was approximately twice the value obtained by the Oxford group for the same, $\lambda = 648$ nm, transition. The Moscow group had originally reported (Bogdanov et al. 1980a) a value for R in disagreement with both the theoretical predictions and with the experimental results of both the Oxford and Novosibirsk groups. Their value was $R_{exp}/R_{theor} = -0.02 \pm 0.1$. A second publication (Bogdanov et al. 1980b) reported a value of $R = (-2.4 \pm 1.3) \times 10^{-8}$, still in disagreement with both theory and the other experimental measurements. They noted, however, that the errors within an individual series of measurements (see Table 8.2) exceeded the standard deviation in some cases, indicating that there were additional systematic errors present that varied comparatively slowly with time. The Moscow group continued their investigation of the sources and magnitudes of these systematic errors and found two principal sources of difficulty: a change in the spatial distribution of light intensity as the frequency of the laser was changed and interference between the signal beam and scattered light in the experimental apparatus (Birich et al. 1984). They took steps to minimize these effects and to measure any residual effects, and noted that their earlier results had not included all of these controls and corrections. Their final value was $R = (-7.8 \pm 1.8) \times 10^{-8}$ and they concluded, "It is clear that our latest results and the results of the Oxford group [the Oxford group was reporting a value of approximately $(-9 \pm 2) \times 10^{-8}$ at this time] are in sufficient agreement with one another and with the results of the most detailed calculations" (Birich et al. 1984, p. 448).

The Oxford group has continued their measurements on the $\lambda = 648$ nm line in bismuth up to the present (Taylor et al. 1987). They had presented intermediate reports at conferences in 1982 and 1984 consistent with the W-S theory. Their most recent result $R = (-9.3 \pm 1.4) \times 10^{-8}$ is consistent with the Standard Model (the uncertainty is now primarily in the theoretical calculations, as we have seen) and with the measurements of Birich et al. (1984), but inconsistent with those of Barkov and Zolotorev (1980). During the 1980s the group has devoted considerable effort to searching for systematic effects and trying to eliminate or correct for them. The difficulties of this type of experiment are severe. They note that their method depends on changing the wavelength of the laser and that wavelength-dependent angles (WDA), comparable to the expected parity-nonconserving optical-rotation angle, are seen in their apparatus, even in the absence of bismuth.[21] This WDA varied with time in an apparently random way and affected the group's ability to make measurements and carry out diagnostic tests. They do not expect the WDA to give rise to systematic error in the bismuth measurements, because of its random nature, but its presence does indicate the possibility of angle effects of similar size to the expected effect and the need for precautions and checks. Their paper lists the following experimental checks: (a) angle sensitivity, (b) angle lock, (c) polarizer reversal, (d) Faraday contamination, (e) pickup, (f) cross modulation between laser and magnetic field, (g) transverse magnetic field effects, and (h) oven reversal. As

---

[21] Recall that the Moscow group also saw such effects.

Table 8.4.

Experiment	$10^8 R$		Theory
Bi 648 nm			
Novosibirsk (1979)	-20.2 ± 2.7	-13	Sandars (quoted in Barkov and Zolotorev, 1979)
Oxford (1984)	-9.3 ± 1.5	-17	Novikov et al. (1976)
Moscow (1984)	-7.8 ± 1.8	-18.8	Barkov and Zolotorev (1980b)
		-10.5	Martensson, Henley, and Wilets (1981)
Bi 876 nm			
Seattle (1981)	-10.4 ± 1.7	-8	Martensson et al. (1981)
		-11	Sandars (1984)
		-13	Novikov et al. (1976)

one can see, making a valid measurement demands considerable care. The paper notes that their present result disagrees with their earlier published value of $R = (+2.75 \pm 4.7) \times 10^{-8}$. Because their new result involves an improved apparatus, considerably more data, and numerous checks against possible systematic error, they prefer their latest result. They conclude that their earlier result was in error. They admit, however, that they do not have any explanation of what the error was and the difference between the measurements. The rebuilding of the apparatus precludes testing many of the likely explanations.

The present situation is virtually the same as when M. Bouchiat and Pottier (1984) presented their summary, Table 8.4. The bismuth results are in approximate agreement with the Weinberg-Salam theory, although the discrepancy with the Novosibirsk measurement remains slightly worrisome. S. Blundell has told me that there are recent reports of a new Novosibirsk experiment whose results agree with those of Oxford (private communication).

Atomic parity violations have also been observed in elements other than bismuth. We mentioned earlier an experiment on thallium (Conti et al. 1979), which had given a result in approximate agreement with the W-S theory. This experiment was part of the context in which the early bismuth experiments had been evaluated. The experiment had measured the circular dichroism, δ, and had found $\delta = (+5.2 \pm 2.4) \times 10^{-3}$ in agreement with the theoretical value $\delta_{th} = (+2.3 \pm 0.9) \times 10^{-3}$. Commins and his collaborators continued this series of experiments through the 1980s using the same basic method, although they made improvements in the experimental apparatus and carried out more thorough investigations of possible sources of systematic error. In 1981 (Bucksbaum, Commins, and Hunter 1981a, 1981b), they reported a value $\delta = (+2.8 \pm \substack{1.0 \\ 0.9}) \times 10^{-3}$ in comparison with the theoretical value $(+2.1 \pm 0.7) \times 10^{-3}$. The change in the theoretical value of δ was caused by a change in the value of $\sin^2\theta_W$, an experimentally measured parameter in the Weinberg-Salam theory, from 0.25 to 0.23. In this paper the comparison between theory and experiment was also expressed in terms of Q_W, the weak charge defined in the W-S theory as $Q_W = Z(1 - \sin^2\theta_W) - N$, where Z is the nuclear charge and N the number of neutrons. For thallium $Q_W(^{205}_{81}Tl) = -117.5$ and $Q_W(^{203}_{81}Tl) = -115.5$, for $\sin^2\theta_W = 0.23$. They found $Q_{W,\exp} = -155 \pm 63$, where the uncertainty included both the experimental uncertainty in δ as well as the uncertainty in the theoretical quantities needed to calculate Q_W from δ. This experiment also included systematic corrections due to imperfect circular polarization, misaligned electric fields, and residual magnetic fields that were "determined precisely by a series of auxiliary experiments." These experimental results agree with the theoretical predictions, although the uncertainty is rather large.

The most recent measurement on thallium (Drell and Commins 1985) gave a value of $Q_{W,\exp}(^{205}_{81}Tl) = -188 \pm 47$ in comparison with a theoretical prediction of -113, for $\sin^2\theta_W = 0.215$.[22] This difference, $\Delta Q = Q_{W,\exp} - Q_{W,th} = -75 \pm 47$, was somewhat worrisome for the experimenters. They noted that the uncertainty in $Q_{W,\exp}$ includes both the experimental uncertainty of 19% combined with an uncertainty in the theoretical calculations of 15%. They concluded, "One can speculate on alternatives to the standard model that would lead to $\Delta Q \neq 0$. For $^{205}_{81}Tl$, we believe that such speculations are premature and must remain so until β [a quantity that had only been calculated] is accurately measured and the RMBPT calculation is thoroughly checked" (Drell and Commins 1985, p. 2209).

---

[22] The accepted value of $\sin^2\theta_W = 0.226$. This changes the value of Q_W to -116. This does not change any of the conclusions.

That direct measurement of β has been performed by Tanner and Commins (1986). Using measurements on Stark-induced amplitudes (in an electric field) they found $\beta = (1.09 \pm 0.05) \times 10^{-5} \mu_0$ cm/V, where μ_0 is the electron Bohr magneton (a measured quantity). This is in contrast to the calculated value used by Drell and Commins (Neuffer and Commins 1977) of $\beta = (1.64 \pm 0.25) \times 10^{-5}\mu_0$ cm/V in their determination of $Q_{W,exp}$ for thallium. Using the measured value of β gives a value of $Q_{W,exp}$ $\binom{205}{81}$Tl$) = -125 \pm 25$, in contrast to the previously reported value of -188 ± 47.[23] The new value of $\Delta Q = -12 \pm 25$ shows the good agreement of experiment with theory.[24]

Parity-nonconserving optical rotation has also been observed in lead by the Washington group (Emmons, Reeves, and Fortson 1983). Their experimental value of $R = (-9.9 \pm 2.5) \times 10^{-8}$ agrees, to within the uncertainties of both the measurement and the atomic-theory calculation, with the theoretical prediction of $R = -13 \times 10^{-8}$.[25]

A series of measurements has also been done on cesium. As early as 1974, even before the existence of the weak neutral currents predicted by the Weinberg-Salam theory had been established, Bouchiat and Bouchiat (1974) had calculated the expected effect of such neutral currents in atomic parity-violating experiments. They had found that the effect would be enhanced in heavy atoms: "Going from hydrogen to cesium, one gets an enhancement of the order of 10^6."

---

[23] The uncertainty in Q_W is reduced because the 5% uncertainty in the measured value of β is smaller than the 15% uncertainty in the calculated value.

[24] This direct measurement of β has not, however, eliminated all of the uncertainty concerning the comparison between experiment and theory. A recent calculation by H. Zhou (private communication, 1988) gives a value of β for thallium $= 1.56 \times 10^{-5} \mu_0$ cm/V. This is close to the value calculated by Neuffer and Commins (1977) and used by Drell and Commins (1985) of $\beta = 1.64 \times 10^{-5} \mu_0$ cm/V. Zhou's value of β gives a value of $Q_{W,exp}$ (Tl) $= -166 \pm 37$ and $\Delta Q = -66 \pm 37$. Using thallium parameters calculated by Johnson et al. (1985, 1986), Zhou finds $\beta = 1.18 \times 10^{-5} \mu_0$ cm/V if she includes core polarization, as experiment suggests. This last value is in reasonable agreement with the measured value, $\beta = (1.09 \pm 0.05) \times 10^{-5} \mu_0$ cm/V. I should emphasize here that the uncertainty in Zhou's calculation is itself uncertain. Clearly, some uncertainty remains in the comparison of theory and experiment.

[25] This is their average of two different calculations, which gave -11×10^{-8} and -14×10^{-8}, respectively (Novikov, Sushkov, and Khriplovich 1976; Botham, Martensson, and Sandars 1981).

The first experimental result on cesium was reported by M. Bouchiat and her collaborators (M. Bouchiat et al. 1982). They found that the parity-nonconserving parameter $Im(E_1^{PNC}/\beta)_{exp} = (-1.34 \pm 0.22 \pm 0.11)$ mV/cm, where the theoretical value was (-1.73 ± 0.07) mV/cm. They concluded, "In view of the experimental and theoretical uncertainties, this is quite consistent with the measured value" (p. 369). This measurement was on a $\Delta F = 0$ hyperfine transition. A second paper (M. Bouchiat et al. 1984) reported a measurement on a $\Delta F = 1$ transition in cesium and found $Im(E_1^{PNC}/\beta) = (-1.78 \pm 0.26 \pm 0.12)$ mV/cm. "Within the quoted uncertainties, the two results clearly agree, so the two measurements successfully cross-check one another. It is then fair to combine them, which yields $Im(E_1^{PNC}/\beta) = -1.56 \pm 0.17 \pm 0.12$ mV/cm" (p. 467). The theoretical value had changed slightly and was $(-1.61 \pm 0.07 \pm 0.20)$, so theory and experiment were in agreement.

The latest experiment on cesium has been performed by Carl Wieman and his collaborators (Gilbert et al. 1985; Gilbert and Wieman 1986; Wieman, Gilbert, and Noecker 1987). They found $Im(E_1^{PNC}/\beta) = -1.65 \pm 0.13$ mV/cm, in good agreement with the previous measurement by M. Bouchiat et al. and with the theoretical prediction. Their value of $Q_{W,exp}(\text{Cs}) = -77 \pm 6 \pm 1.5$ is in agreement with the theoretical value of $-71 \pm 1.7 \pm 3.0$. This is the first atomic parity violation experiment to obtain an uncertainty of less than 10 percent. The experimental checks were extensive. They included four independent spatial reversals of experimental conditions to identify the parity-nonconserving signal when, in principle, only two are required to resolve the effect. This reduced the potential systematic error because nearly all the factors that can affect the transition rate are correlated with, at most, one of these reversals. Other possible sources of systematic error were identified and their possible effects measured in auxiliary experiments. The experimenters also introduced known nonreversing fields, misalignments, and so forth. The measured effect of these interventions agreed with their calculations of these effects and also indicated that these effects were small compared to the parity-violating signal. Their analysis of their data over time scales from minutes to days also indicated that the distribution of their measured values of $Im(E_1^{PNC}/\beta)$ was completely statistical and that time-dependent systematic effects were small.

In addition, as mentioned above, atomic parity-violation experiments test the W-S theory in a very different energy range and test different electron–quark couplings from the high-energy physics experiments. They also provide, as will be discussed subsequently, a very severe test of the theory itself. From the experimental result one can calculate the axial vector electron–proton coupling constant. They found $C_{2p} = -2 \pm 2$, in good agreement with the theoretical value of 0.1, and a substantial improvement over the previously reported experimental limit of < 100 (M. Bouchiat et al. 1984). The experimental result gives a value of $\sin^2\theta_W = 0.254 \pm 0.028 \pm 0.007$. "Although this is a respectable determination of θ_W the uncertainty is several times larger than that of the best high-energy determinations" (Wieman et al. 1987, p. 73). (The best high-energy result is 0.226 ± 0.004.)

This experiment also illustrates the value of repeating measurements of the same quantity with increasing precision and accuracy because of the possible future utility of testing either existing theory or further theoretical developments. When these experiments were first proposed in 1974 their purpose was to test the Weinberg-Salam theory in a different energy range and in a different physical system. Developments since then have given them another purpose (C. Bouchiat and Piketty 1983; Robinette and Rosner 1982).

However, the primary significance of atomic PNC measurements such as this is not in their determination of this free parameter of the standard model [$\sin^2\theta_W$] but rather in the way they test the validity of the model itself. This can be seen in Figure 8.4. This is a plot showing the experimentally allowed values of the electron–quark coupling constants C_{1d} and C_{1u}. If the standard model is a perfect description of nature, the correct values for these constants must lie along the line $SU_2 \times U_1$. Atomic PNC measurements restrict the possible values to a band which is nearly parallel to this line. Thus they are relatively poor at determining exactly which point on the line is correct (i.e. the value of $\sin^2\theta_W$). However, this makes them uniquely sensitive to deviations off the line which can happen if the standard model is not totally correct. For comparison the best high energy restrictions in this plane are from the SLAC

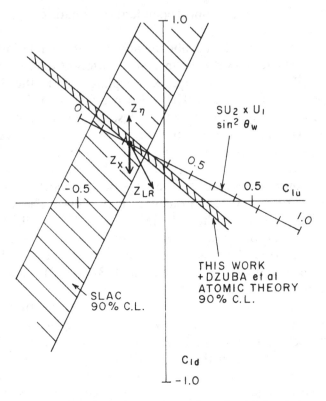

Figure 8.4. Experimental constraints on the parity-violating electron–quark coupling constants C_{1d} and C_{1u}. The arrows show the effects of additional neutral bosons of three different types. From Wieman, Gilbert, and Nœcker (1987).

electron scattering experiments and one can see it is very insensitive to this type of deviation from the standard model line. A number of the most attractive alternatives to the standard model have additional neutral bosons which would lead to exactly this sort of deviation and for these cases atomic PNC measurements provide important, and often the best, experimental constraints on the parameters of these models. To illustrate this we have shown the direction of deviation one would get for three models which have additional neutral bosons and have been studied extensively. These are the Z_χ, which occurs in SO_{10} models of grand unification; the Z_{LR}, which is found in many left–right symmetric theories; and the Z_η, which occurs in many superstring theories. By restricting the possible values of C_{1d} and C_{1u} the present experiment sets lower limits on the masses of these neutral bosons (Wieman et al. 1987, pp. 73–5).

It is fair to say that the current situation with respect to atomic parity-violation experiments and the Weinberg-Salam theory is that the preponderance of evidence favors the theory.

8.3 DISCUSSION

There are several points worth making about this episode of atomic parity-violation experiments. Perhaps most important is that the comparison between experiment and theory can often be extremely difficult. This is particularly true when, as in this episode, one is at the limit of what one can calculate confidently and what one can measure reliably.

In 1977, the atomic physics calculations of the expected parity-violating effects had a large uncertainty. Even today, when estimates of the accuracy of the calculations are about 25 percent (Drell and Commins 1985; C. Tanner, private communication) and the ratio of the highest to lowest predicted values of the effect in bismuth is 1.8 (see Table 8.4). Zhou's calculated values of β for thallium, discussed earlier, vary from $1.18 \times 10^{-5} \ \mu_0$ to $1.56 \times 10^{-5} \ \mu_0$, depending on the assumptions used and the approximations made.

The experimental results are also uncertain.[26] These experiments contain systematic errors, which may mimic or mask the expected result. Unlike statistical errors, which can be calculated precisely, systematic errors are both extremely difficult to detect and to estimate. We have seen that experimenters have devoted enormous effort to find such systematic errors and to eliminate them. Sometimes they have been measured and corrected for. Some are still unknown.

Pickering remarks that by 1979, and presumably to this day, there had been no *intrinsic* change in the early Washington and Oxford results. In the sense that no one knows with certainty why those early results were wrong, he is correct. Nevertheless, since those early experiments physicists have found new sources of systematic error that were not dealt with in the early experiments. The redesign of the apparatus has, in many cases, precluded testing whether or not these effects were sig-

---

[26] The smallest uncertainty in any experimental result is 10% (Gilbert et al. 1985; Gilbert and Wieman 1986; Wieman et al. 1987).

nificant in the older apparatus. Though one cannot claim with certainty that these effects account for the earlier, presumably incorrect, results, one does have reasonable grounds for believing that the later results are more accurate. The consistency of the later measurements, especially those done by different groups, enhances that belief.

It is clear that the relationship between theory and experiment is more complex than "Man proposes, Nature disposes." Our history has shown that it is not at all clear at times just what man is proposing. The experimental results show that one may not be able to see just how Nature is disposing.

What does seem clear, however, is that the evidence model fits the history of this episode better than Pickering's model. Scientists *chose*, on the basis of reliable experimental evidence provided by the SLAC E122 experiment, to accept the Weinberg-Salam theory. They *chose* to leave an apparent, but also quite uncertain, anomaly in the atomic parity-violation experiments for future investigation.[27] I will have a general discussion of the differences between my view of science and that of the social constructivists in the concluding chapter.

This episode also demonstrates that scientists make judgments about the reliability of experimental results that coincide with what one would decide on epistemological grounds. The E122 group argued for the validity of their experimental result using strategies that coincide with an epistemology of experiment (see note 19). As we have seen, the scientific community accepted their arguments.

We have also seen the utility of repeating experimental measurements, albeit with more precision and accuracy, because of possible future theoretical developments. When these atomic parity-violation experiments began, their purpose was to demonstrate the existence of parity-violating effects. Subsequent theoretical work enables one to use these experiments as stringent tests of the model and of proposed alternatives. Current plans to increase the precision of the cesium measurements to 1 percent will provide even more severe tests (C. Tanner, private communication).

[27] As we have seen, the future investigation removed the anomaly.

Conclusion

The title of this book might suggest an unthinking adherence to the roles of experiment in science and an unquestioned faith in experimental results. This is not the conclusion I want the reader to reach. At the very least, the reader should firmly believe in both the fallibility and the corrigibility of science. I also want the reader to be persuaded that the evidence model, in which questions of theory choice, confirmation of theory, or refutation of theory are decided on the basis of valid experimental evidence, is both an accurate description of the way science is practiced, and a philosophically justified view.

There are episodes in which the experimental results are still unchallenged and regarded as correct. Thus, the experiments of Wu, Ambler, Hayward, Hoppes, and Hudson; of Garwin, Lederman, and Weinrich; and of Friedman and Telegdi[1] provided the crucial evidence that established the hypothesis of parity nonconservation, suggested by Lee and Yang. The discovery of parity nonconservation played an essential role in the development of the V-A theory of weak interactions. It was also incorporated into the Weinberg-Salam theory the successor to the V-A theory. The experiments on atomic parity violation were regarded as extremely important, if not crucial, tests of that theory. I suspect that had the results not agreed with the Weinberg-Salam theory

---

[1] I have included the names of all the experimenters, at least once, because there has been a tendency to underestimate their role in science, even to the extent of omitting their names. Everyone remembers Planck for his work on quantum theory, but few people remember that it was the experimental work of Lummer, Pringsheim, Rubens, and Kurlbaum that provided the experimental basis for that work. This problem has become even more acute in modern high-energy physics experiments in which the number of authors of a paper can sometimes exceed one hundred. Even I would not list them all.

the theory would have been discarded, or at least severely modified.

The interaction between experiment and theory is not always as clear-cut as it was in the case of the discovery of parity non-conservation. The experiment of Christenson, Cronin, Fitch, and Turlay established the existence of $K^{\circ}_2 \rightarrow 2\pi$ decay and thus, of apparent *CP* violation. Further experiments eliminated alternative explanations and established *CP* violation, a hypothesis still regarded as correct. Even in this case, experiment was fallible. At least one of the sets of experimental measurements of η_{+-}, the *CP*-violating parameter, must be wrong. Although no firm conclusion has been reached as to which set is correct most physicists would, I believe, favor the second set, and regard this as an example of the corrigibility of experimental results. I note that although the Princeton experiment called for a new theory, one of the roles of experiment, no acceptable theory has been forthcoming.

The fallibility and corrigibility of experimental results have been further illustrated by the experimental anomalies for the *V-A* theory. When that theory was proposed, no fewer than four different, and seemingly valid, experimental results were in disagreement with its predictions. The experiments were reanalyzed and redone, the results were changed, and the anomalies were removed.

The history of β decay, recounted in the first three chapters, shows that fallibility in science is sometimes even more complex. The early measurements on β-decay spectra refuted Fermi's theory and supported the Konopinski-Uhlenbeck modification. Later, after problems were found due to scattering and energy loss in the sources, the experiments were modified and the new results reversed the decision. The blame does not always fall only on experiment. Fallibility also applies to theoretical calculation and analysis. The analysis of the RaE spectrum by Petschek and Marshak, which established the pseudoscalar (*P*) interaction in β decay, turned out to be incorrect. The later analysis did not require the *P* interaction. The comparison between experiment and theory may also be fallible. The early experimental results on forbidden transitions were compared to the theoretical predictions for allowed transitions. It is not surprising that they disagreed. When the calculations of the forbidden spectra were done, the discrepancy disappeared.

The fallibility and corrigibility of both theory and experiment is illustrated, once again, in the history of atomic parity-violation experiments discussed in the last chapter.

Does this demonstrated fallibility of experiment, of theoretical calculation, and of the comparison between experiment and theory preclude a legitimate role for experiment in theory choice, confirmation, or refutation? I think not. The history presented shows that although the fallibility is worrisome, corrigibility also exists. At any given time, the use of the best established experimental results is reasonable. I know of no better basis on which to decide these questions.[2] This may, occasionally lead us to incorrect decisions, but that is a risk we have to take.

I believe that the history given also shows that the evidence model of science is a good description of the way science is done. I also hope that the arguments given have persuaded the reader that this model is philosophically legitimate. I have presented arguments in favor of my proposed epistemology of experiment. This set of strategies establishes the validity of experimental results, fallible though they may be. These results are then reasonably used in theory choice, confirmation, and refutation.

I also hope that I have persuaded the reader that the Bayesian approach to the philosophy of science is a fruitful way of looking at science. It provides adequate descriptions of historical episodes and justification for many of our intuitions about the conduct of good science, such as the epistemology of experiment and the use of varied evidence. It also requires that observation of evidence entailed, or predicted, by a hypothesis should strengthen our belief in that hypothesis.

Can the comparative analysis of one episode, that of atomic parity-violation experiments, or two, if the reader accepts my judgment about the analyses of the weak neutral-current experiments, establish the superiority of the evidence model over the social constructivist view? I believe so, because the evidence model is also supported by other case studies from the history of physics.[3] If the social constructivist view were correct then one would expect to find at least one episode in which the decision of the scientific community went against the weight of experimental evidence. No such episode has been provided. I believe

---

[2] The Bayesian view argues that this is correct.
[3] See, for example, the cases discussed in Galison (1987).

that the best case presented so far has been Pickering's analysis of the atomic parity-violation experiments, which I have just discussed in detail. In this case, Pickering claims that the decision on theory choice excluded experimental evidence of equal weight that argued against that choice. I have argued that Pickering's analysis is incorrect.

Nevertheless, I believe that the social constructivists are right in insisting that in some cases the cognitive interests ("theoretical presuppositions") of the scientists do play a role in establishing experimental results, and thus in theory choice.[4] Galison's study of gyromagnetic experiments in the early twentieth century (1987, ch. 2) pointed out that theory often provides experimenters with quantitative predictions that enable them to find the effects sought or to separate these effects from background sources of error. Such predictions may also influence the decision to stop looking for sources of error, declare the experiment ended, and report the result, which may be that predicted by theory. In the early experiments it seems that the theoretical presuppositions of the experimenters did lead to incorrect results, or at least to results that disagree with currently accepted results. They were, however, in agreement with the existing theoretical predictions. Similarly, Millikan excluded at least one event from his paper on the oil-drop experiment because it did not agree with his presupposition that electric charge was quantized (Franklin 1986, ch. 5). These are cases in which the theoretical presuppositions of the experimenters led to incorrect results. In both cases the importance of the experiments led to many repetitions and to the correction of these early results. Changes in the evidential context of the weak neutral-current experiments, that is, the new, well-corroborated Weinberg-Salam theory, led scientists to look for such currents and to reinterpret earlier experimental results.

Suppose, however, someone did present a case in which the decision went against the experimental evidence. Would this destroy the evidence model? I don't think so. I believe that the evidence model describes not only what scientists should do, but

---

[4] The devoted reader, who has read my 1986 book, will notice that I have modified my position on this issue. I did not believe that in the episodes discussed there the presuppositions of the experimenters were important. I still think that is true for those episodes, but I now believe that theoretical presuppositions can be a source of fallibility.

also what they, in fact, do most of the time. That scientists, being human, are fallible and do not always behave as they ought to should surprise no one. Of course, if scientists' decisions went against the weight of evidence most of the time I would, on the basis of the evidence model, give up the evidence model as an adequate description of scientific practice. I would still argue, however, that it is the way science should be done.

Science is a social construction because it is constructed by the scientific community. But as both this study and other studies demonstrate, it is constructed from experimental evidence, rational discussion, criticism, and the inventiveness of scientists. As Karl Kautsky (1902, p. 13) remarked, "The fact that an idea emanates from any particular class, or accords with their interests, of course proves nothing as to the idea's truth or falsity."

Appendix

We may clarify this idea of an "important" confirming or refuting experiment, and the dependence of this importance on the theoretical context using Bayesian confirmation theory. Let D be the Dirac theory and e_1 be the existence of the positron. Applying Bayes's Theorem, we get

$$P(D|e_1) = P(e_1|D) \, P(D)/P(e_1),$$

where $P(e_1|D) = 1$ because $D \vdash e_1$. We can also write

$$P(e_1) = P(e_1|D) \, P(D) + P(e_1|\neg D) \, P(\neg D),$$

where $\neg D$ is the negation of D. Because no other theory, at the time, predicted the existence of the positron, we can set $P(e_1|\neg D) \, P(\neg D) = \beta$, where β is very small.

Thus, $P(D|e_1) = P(D)/(P(D)+\beta) \approx 1 - \epsilon \approx 1$, where $\epsilon = \beta/P(D)$ is also quite small. This confirms our intuition that observation of a result predicted by only one theory gives us very good reason to believe that theory. This is not, of course, the only type of important experiment. The examples discussed earlier of the experiments on parity nonconservation and CP violation provide an illustration of another type. These are experiments that eliminate competing hypotheses or theories while confirming one of the alternatives. These also provide substantial confirmation. We now set $P(D|e_1) = P_1(D)$, the new prior probability of D. Let us now consider that happens to our belief in D when e_2, the symmetric double scattering of electrons, is observed.

$$P(D|e_2) = P(e_2|D) \, P_1(D)/P(e_2). \tag{1}$$

$$P(e_2) = P(e_2|D) \, P_1(D) + P(e_2|\neg D) \, P_1(\neg D). \tag{2}$$

Let $K = P(e_2|\neg D)/P(e_2|D)$; then, from equation (2)

$$P(e_2) = P(e_2|D) \, P_1(D) + K \, P(e_2|D) \, [1 - P_1(D)]$$

but $P_1(D) = 1 - \epsilon$,

$$P(e_2) = P(e_2|D) \, [1-\epsilon] + K \, P(e_2|D) \, [1 - 1 + \epsilon]$$

$$P(e_2) = P(e_2|D) \, [1 - \epsilon + K\epsilon].$$

We can neglect ϵ, which is very small; so, from equation (1),

$$P(D|e_2) = P(e_2|D) \, P_1(D)/P(e_2|D) \, [1+K\epsilon] = P_1(D)/1 + K\epsilon.$$

We expect $K \leq 1$, because $\neg D$ is a weak statement compared with D and in this case we see $P(D|e_2) \approx P_1(D)$, that is, our confidence in D remains virtually unchanged. The result is quite robust for differing values of K, depending on ϵ. Thus if $\epsilon = 0.01$, K can be as large as ten without changing our belief in D by more than 10 percent.

Let us now see what we can say about $P(H|e_2)$, where H represents the auxiliary hypotheses used by Mott to derive $\neg e_2$, the asymmetric scattering: single, large angle scattering of high-velocity electrons from a heavy nucleus. By a theorem of the probability calculus,

$$P(H|e_2) + P(D|e_2) - P(H \wedge D|e_2) = P(H \vee D|e_2).$$

But $P(H \wedge D|e_2) = 0$, because $H \wedge D \neg e_2$; so

$$P(H|e_2) = P(H \vee D|e_2) - P(D|e_2). \tag{3}$$

But $P(H \vee D|e_2) \geq P(D|e_2)$, so we can write $P(H \vee D|e_2)$ as $1 - \delta$, where $\delta \leq K\epsilon$.

$$P(H|e_2) = 1 - \delta - (1 - K\epsilon) = K\epsilon - \delta,$$

which is very small.

Thus, as both our simple calculation and the history show, our belief in H becomes very low, whereas our belief in D remains virtually the same. Scientists did in fact offer alternatives to H.

This is a solution to a very particular case of the Duhem-Quine problem in which $P(D)$ is originally very large. In the general case, in which $P(D|e_2)$ is not very close to 1, we cannot say anything, as equation (3) shows. It does indicate, however, that when $P(H|e_2)$ is close to 1, D will receive the blame, as we expect.

This also demonstrates the fruitfulness and utility of the Bayesian approach. It can explicate, quite naturally, this episode of Dirac theory, which is otherwise explicable only in the vaguest sense.

References

ABBREVIATIONS

AJP	*American Journal of Physics*
CR	Académie des Sciences, Paris
	Comptes rendus hebdomaires des séances
JETP	*Zhurnal Eksperimental 'noi i*
	Teoreticheskoi Fiziki
NC	*Il Nuovo Cimento*
Phil. Mag.	*Philosophical Magazine*
PR	*Physical Review*
PRL	*Physical Review Letters*
PL	*Physics Letters*
Proc. Roy. Soc.	*Proceedings of the Royal Society (London)*
RMP	*Reviews of Modern Physics*
Z. Phys.	*Zeitschrift für Physik*

Abashian, A. et al. 1957. "Angular Distribution of Positrons Observed in a Liquid Hydrogen Bubble Chamber," *PR* 105: 1927–8.

Achinstein, P., and Hannaway, O. 1985. *Observation, Experiment, and Hypothesis in Modern Physical Science.* Cambridge, MA: MIT Press.

Ackermann, R. 1985. *Data, Instruments, and Theory.* Princeton, NJ: Princeton University Press.

    1989. "The New Experimentalism," *British Journal for the Philosophy of Science,* 40: 185–90.

Adams, R. V. et al. 1948. "Cosmic Rays at 30,000 Feet," *RMP* 20: 334–49.

Ahrens, T.; Feenberg, E.; and Primakoff, H. 1952. "Pseudoscalar Interaction in the Theory of Beta-Decay," *PR* 87: 663–4.

Albert, R. D., and Wu, C. S. 1948. "The Beta-Spectrum of S^{35}," *PR* 74: 847–8.

Alburger, D. E. 1950. "Beta-Ray Spectrum of K^{40}," *PR* 79: 236.

Alburger, D. E.; Hughes, D. J.; and Eggler, C. 1950. "The β-Spectrum of Be^{10}," *PR* 78: 318.

Alford, W. P., and Hamilton, D. R. 1954. "Recoil Spectrum in the β Decay of Ne^{19}," *PR* 94: 779.

    1955. "Recoil Spectrum in the Beta Decay of Ne^{19}," *PR* 95: 1351–3.

1957. "Electron–Neutrino Angular Correlation in the Beta Decay of Ne^{19}," *PR* 105: 673–8.

Alichanian, A. I.; Alichanow, A. I.; and Dzelepow, B. S. 1938. "On the Form of the β-Spectrum of RaE in the Vicinity of the Upper Limit and the Mass of the Neutrino," *PR* 53: 766–7.

Alichanian, A. I., and Nikitin, S. J. 1938. "The Shape of the β-Spectrum of ThC and the Mass of the Neutrino," *PR* 53: 767.

Alichanow, A. I.; Alichanian, A. I.; and Dzelepow, B. S. 1936. "The Continuous Spectra of RaE and P^{30}," *Nature* 137: 314–15.

Allen, J. S. 1942. "Experimental Evidence for the Existence of a Neutrino," *PR* 61: 692–7.

Allen, J. S., and Jentschke, W. K. 1952. "Electron–Neutrino Angular Correlation in the Beta-Decay of He^6," *Bulletin of the American Physical Society* 2: 17.

1953. "Electron–Neutrino Angular Correlation in the Beta-Decay of He^6," *PR* 89: 902.

Allen, J. S.; Paneth, H. R.; and Morrish, A. H. 1949. "Electron–Neutrino Angular Correlation in the Beta-decay of He^6," *PR* 75: 570–7.

Alvarez, L. W. et al. 1950. "Electrical Detection of Artificially Produced Mesons," *PR* 77: 752.

Ambler, E. et al. 1957. "Further Experiments on β Decay of Polarized Nuclei," *PR* 106: 1361–6.

Ambrosen, J. 1934. "Uber den aktiven Phosphor und des Energiesspektrum seiner β-Strehlen," *Z. Phys.* 91: 43–8.

Anderson, C. D., and Neddermeyer, S. H. 1936. "Cloud Chamber Observation of Cosmic-Rays at 4300 Meters Elevation and Near Sea Level," *PR* 50: 263–71.

Anderson, C. D. et al. 1947. "On the Mass and the Disintegration Products of the Mesotron," *PR* 72: 724–7.

Anderson, H. L., and Lattes, C. M. G. 1957. "Search for the Electronic Decay of the Positive Pion," *NC* 6: 1356–81.

Arnison, G. et al. (UA1 Collaboration). 1983. "Experimental Observation of Isolated Large Transverse Energy Electrons with Associated Missing Energy at $\sqrt{s} = 540$ GeV," *PL* 122B: 103–16.

(UA1 Collaboration). 1984. "Experimental Observation of Events with Large Missing Transverse Energy Accompanied by a Jet of Photon(s) in pp̄ Collisions at $\sqrt{s} = 540$ GeV," *PL* 139B: 115–25.

Ascoli, G. et al. (eds.). 1958. *High Energy Nuclear Physics, Proceedings of the Seventh Rochester Conference.* New York: Interscience.

Auger, P.; Maze, R.; and Chaminade, R. 1941. "Une démonstration directe de la désintégration spontanée du méson," *CR* 213: 381–3.

Backus, J. 1945. "The Beta-Ray Spectra of Cu^{64} at Low Energies," *PR* 68: 59–63.

Baird, P. E. G. et al. 1976. "Search for Parity-Nonconserving Optical Rotation in Atomic Bismuth," *Nature* 264: 528–9.

1977. "Search for Parity-Nonconserving Optical Rotation in Atomic Bismuth," *PRL* 39: 798–801.

Ballam, J. et al. (eds.). 1957. *High Energy Nuclear Physics, Proceedings of Sixth Annual Rochester Conference*. New York: Interscience.

Banner, M. et al. (UA2 Collaboration). 1983. "Observation of Single Isolated Electrons of High Transverse Momentum in Events with Missing Transverse Energy at the CERN pp̄ Collider," *Physics Letters* 122B: 476–85.

Barkov, L. M., and Zolotorev, M. S. 1978a. "Observations of Parity Nonconservation in Atomic Transitions," *JETP Letters* 27: 357–61.

1978b. "Measurement of Optical Activity of Bismuth Vapor," *JETP Letters* 28: 503–6.

1979. "Parity Violation in Atomic Bismuth," *PL* 85B: 308–13.

1980a. "Parity Violation in Bismuth: Experiment," in Williams (1980), 52–76.

1980b. "Parity Nonconservation in Bismuth Atoms and Neutral Weak-Interaction Currents," *JETP* 52: 360–9.

Barut, A. 1980. "Stable Particles as Building Blocks of Matter," *Surveys in High Energy Physics* 1: 113–40.

1982. "Composite Models of Pion and Muon and a Calculation of the Fermi Coupling Constant from Their Decay," *Kinam* 4: 151–64.

Batens, D., and van Bendegem, J. P. 1988. *Theory and Experiment*. Dordrecht: Reidel.

Beck, G., and Sitte, K. 1933. "Zur theorie der β-Zerfalls," *Z. Phys.* 86: 105–19.

Bell, J. S., and Perring, J. 1964. "2π Decay of the K°_2 Meson," *PRL* 13: 348–9.

Bell, P. R., and Cassidy, J. M. 1950. "Beta-Spectrum of Be^{10}," *PR* 77: 301.

Bell, P. R.; Weaver, B.; and Cassidy, J. M. 1950. "The Beta-Rays of K^{40}," *PR* 77: 399–400.

Bellamy, E. H., and Moorhouse, R. G. (eds.). 1955. *Proceedings of the 1954 Glasgow Conference on Nuclear and Meson Physics*. Oxford: Pergamon Press.

Benvenuti, A. et al. 1976. "Evidence for Parity Nonconservation in the Weak Neutral Current," *PRL* 37: 1039–42.

Bernstein, J. 1967. *A Comprehensible World*. New York: Random House.

Bernstein, J.; Cabibbo, N.; and Lee, T. D. 1964. "*CP* Invariance and the 2π Decay of the K°_2," *PL* 12: 146–8.

Bethe, H. A. 1940a. "The Meson Theory of Nuclear Forces," *PR* 57: 260–72. 1940b. "On the Theory of Meson Decay," *PR* 57: 998–1006.

Bethe, H. A., and Bacher, R. F. 1936. "Nuclear Physics," *RMP* 8: 82–229.

Bielein, H.; Fleischmann, R.; and Wegener, H. 1958. "Polarization of Electrons Emitted by Co^{60}," in Lipkin (1958), 406–7.

Binder, D. 1949. "Analysis of Delayed Coincidence Counting Experiments," *PR* 76: 856–8.

Birich, G. N. et al. 1984. "Nonconservation of Parity in Atomic Bismuth," *JETP* 60: 442–9.

Boehm, F. et al. 1950. "Nuclear Gamma-Radiation of Cu^{61}," *PR* 77, 295–6.

Boehm, F., and Wapstra, A. H. 1957. "Beta-Gamma Circular Polarization Correlation Measurements," *PR* 106: 1364–6.

1958. "Beta-Gamma Circular Polarization Measurements," *PR* 109: 456–61.

Bogdanov, Y. V. et al. 1980a. "Investigation of the Optical Activity of Bi Vapors," *JETP Letters* 31: 214–19.

1980b. "Parity Nonconservation in Atomic Bismuth," *JETP Letters* 31: 522–6.

Bogen J., and Woodward, J. 1988. "Saving the Phenomena," *The Philosophical Review* 97: 303–52.

Bohm, A. et al. 1967. "Observation of Time Dependent K_S and K_L Interference in the $\pi^+\pi^-$ Decay Channel from an Initial K° State," *PL* 27B: 321–7.

Bonetti, A. et al. 1956. "Lo spettro di energie degli elettroni di decadimento dei mesoni μ in emulsione nucleare," *NC* 3: 33–50.

Bonner, T. W. et al. 1936. "Mass of the Neutrino from the Disintegration of Carbon by Deuterons," *PR* 49: 203–4.

Botham, C. P.; Martensson, A. M.; and Sandars, P. G. H. 1981. *Proceedings of the Seventh Vavilov Conference*, ed. S. G. Rautian, 37. Novosibirsk, USSR: Academy of Sciences.

Bouchiat, C. 1980. "Neutral Current Interactions in Atoms," in Williams (1980), 357–69.

Bouchiat, C., and Piketty, C. A. 1983. "Parity Violation in Atomic Cesium and Alternatives to the Standard Model of Electroweak Interactions," *PL* 128B: 73–8.

Bouchiat, M. A., and Bouchiat, C. C. 1974. "Weak Neutral Currents in Atomic Physics," *PL* 48B: 111–14.

Bouchiat, M. A. et al. 1982. "Observation of a Parity Violation in Cesium," *PL*, 117B: 358–64.

1984. "New Observation of a Parity Violation in Cesium," *PL* 134B: 463–8.

Bouchiat, M. A., and Pottier, L. 1984. "Atomic Parity Violation Experiments," in *Atomic Physics 9*, eds. R. Van Dyck and E. Fortson, 246–71. Singapore: World Scientific.

Bowen, D. R. et al. 1967. Measurement of the K^+_{e2} Branching Ratio," *PR* 154: 1314–22.

Bramson, H., and Havens, W. W. 1951. "A Photographic Study of the $\pi^+ \to \mu^+ \to \beta^+$ Decay Process and the Energy Spectrum of the β^+," *PR* 83: 861–2.

Bramson, H. J.; Seifert, A. M.; and Havens, W. W. 1952. "The Energy Spectrum of Positrons from the Decay of the μ Meson," *PR* 88: 304–8.

Brimicombe, M. W. S. M.; Loving, C. E.; and Sandars, P. G. H. 1976. "Calculation of Parity Nonconserving Optical Rotation in Atomic Bismuth," *Journal of Physics B* 9: L237–40.

Bucksbaum, P.; Commins, E.; and Hunter, L. 1981a. "New Observation of Parity Nonconservation in Atomic Thallium," *PRL* 46: 640–3.

1981b. "Observation of Parity Nonconservation in Atomic Thallium," *PR* 24D: 1134–48.

Burgy, M. T. et al. 1957. "Measurement of Beta Asymmetry in the Decay of Polarized Neutrons," *PR* 107: 1731–3.

1958. "Measurement of Asymmetries in the Decay of Polarized Neutrons," *PR* 110: 1214–16.

Caianiello, E. R. 1951. "On the Universal Fermi-Type Interaction," *NC* 8: 749–67.

Callahan, A., and Cline, D. 1965. "Charged $K_{\pi 2}$ Branching Ratio," *PRL* 15: 125–30.

Cartwright, N. 1983. *How the Laws of Physics Lie.* Oxford: Oxford University Press.

Cassels, J. M. et al. 1957. "Radiative Beta Decay of the Pion," *Physical Society of London (Proceedings)* A70: 729–34.

Cavanagh, P. E. 1958a. "Electron–Neutrino Correlation," in Lipkin (1958), 370–4.

1958b. "Electron Polarization," in Lipkin (1958), 394–9.

Cavanagh, P. E. et al. 1957. "On the Longitudinal Polarization of β-particles," *Phil. Mag.* 2: 1105–12.

Christenson, J. H. et al. 1964. "Evidence for the 2π Decay of the K^{0}_{2} Meson," *PRL* 13: 138–40.

Close, F. E. 1976. "Parity Violation in Atoms?" *Nature* 264: 505–6.

Cockcroft, J. D., and Lewis, W. B. 1936. "The Disintegration of Carbon, Nitrogen, and Oxygen," *Proc. Roy. Soc.* A154: 261–79.

Cohen, E. R., and DuMond, J. W. 1965. "Our Knowledge of the Fundamental Constants of Physics and Chemistry," *RMP* 37: 539–94.

Cohen, S. G. et al. 1958. "Measurement of the Circular Polarization of External Bremsstrahlung Produced by β-Particles in the Decay of P^{32}, Y^{90}, and Rh^{106}," in Lipkin (1958), 404–5.

Collins, H. 1976. "The Seven Sexes: A Study in the Sociology of a Phenomenon, or the Replication of Experiments in Physics," *Sociology* 6: 141–84.

Commins, E., and Bucksbaum, P. 1980. "The Parity Nonconserving Electron–Nucleon Interaction," *Annual Reviews of Nuclear and Particle Science* 30: 1–52.

Commins, E. D., and Kusch, P. 1958. "Upper Limit to the Magnetic Moment of He^{6}," *PRL* 1, 208–9.

Conan Doyle, A. 1967. "The Sign of Four," in *The Annotated Sherlock Holmes*, ed. W. S. Baring-Gould. New York: Clarkson N. Potter.

Conti, R. et al. 1979. "Preliminary Observation of Parity Nonconservation in Atomic Thallium," *PRL* 42: 343–6.

Conversi, M.; Pancini, E.; and Piccioni, O. 1947. "On the Disintegration of Negative Mesons," *PR* 71: 209–10.

Conversi, M., and Piccioni, O. 1946. "On the Disintegration of Slow Mesons," *PR* 70: 874–81.

Cook, C. S., and Langer, L. M. 1948. "The Beta-Spectrum of Cu^{64} as a Test of the Fermi Theory," *PR* 73: 601–7.

Cook, C. S.; Langer, L. M.; and Price, H. C. 1948a. "The Beta-Spectrum of S^{35} and the Mass of the Neutrino," *PR* 73: 1395.

1948b. "Study of the Beta-Spectra of C^{14} and S^{35}," *PR* 74: 548–52.

1948c. "Shape of the Positron Spectrum of N^{13}," *PR* 74: 502–3.

Crane, H. R. 1948. "The Energy and Momentum Relations in Beta-Decay, and the Search for the Neutrino," *RMP* 20: 278–95.

Crane, H. R., and Halpern, J. 1939. "Further Developments on the Recoil of the Nucleus in Beta-Decay," *PR* 56: 232–7.

Critchfield, C. L. 1943. "The Antisymmetrical Interaction in Beta-Decay Theory," *PR* 63: 417–25.

Critchfield, C. L., and Wigner, E. P. 1941. "The Antisymmetrical Interaction Beta-Decay Theory," *PR* 60: 412–13.

Cronin, J. W. et al. 1967. "Measurement of the Decay Rate of $K^\circ_2 \rightarrow \pi^\circ\pi^\circ$," *PRL* 18: 25–9.

Crowe, K. M.; Helm, R. H.; and Tautfest, G. W. 1955. "Preliminary Data on the Measurement of the $\mu^+-\beta^+$ Decay Spectrum," *PR* 99: 872–4.

Culligan, G.; Frank, S. G. F.; and Holt, J. R. 1959. "Longitudinal Polarization of the Electrons from the Decay of Unpolarized Positive and Negative Muons," *Physical Society of London (Proceedings)* A73: 169–77.

Culligan, G. et al. 1957. "Longitudinal Polarization of the Positrons from the Decay of Unpolarized Positive Muons," *Nature* 180: 751–2.

Cvijanovich, G. B., and Jeannet, E. A. 1964. "Anisotropie dans la Desintegration $\pi-\mu$ Mesons π^+ Cree dans la Desintegration $K^+ \rightarrow 2\pi^+\pi^-$," *Helvetia Physica Acta* 37: 211–12.

Cvijanovich, G. B.; Jeannet, E. A.; and Sudarshan, E. C. G. 1965. "*CP* Invariance in Weak Interactions and the Pion Decay Asymmetry," *PRL* 14: 117–18.

Davies, J. H.; Lock, W. O.; and Muirhead, H. 1949. "The Decay of μ Mesons," *Phil. Mag.* 40: 1250–60.

DeBouard, X. et al. 1965. "Two Pion Decay of K°_2 at 10 GeV/c," *PL* 15: 58–61.

De Groot, S. R., and Tolhoek, H. A. 1950. "On the Theory of Beta-Radioactivity I: The Use of Linear Combinations of Invariants in the Interaction Hamiltonian," *Physica* 16: 456–80.

De Souza-Santos, M. D. 1942. "On the Decay of Slow Mesons," *PR* 62: 178–9.

Deutsch, M. et al. 1957. "Polarized Positrons from Ga⁶⁶ and Cl³⁴," *PR* 107: 1733–4.

De Waard, H., and Poppema, O. J. 1957. "Longitudinal Polarization of Beta-Particles from ⁶⁰Co, ³²P, and ¹⁷⁰Tm," *Physica* 23: 597–8.

De Waard, H.; Poppema, O. J.; and Van Klinken, J. 1958. "Measurements of the Longitudinal Polarization of Beta-Particles," in Lipkin (1958), 388–93.

Dirac, P. A. M. 1928. "The Quantum Theory of Electrons," *Proc. Roy. Soc.* A117: 610–24.

Drell, P., and Commins, E. 1985. "Parity Nonconservation in Atomic Thallium," *PR* 32A: 2196–210.

Dudziak, W. F.; Sagane, R.; and Vedder, J. 1959. "Positron Spectrum from the Decay of the μ Meson," *PR* 114: 336–58.

Dydak, F. 1979. "Neutral Currents," in *Proceedings of the International Conference on High Energy Physics*, Geneva, June 27–July 4, 1979, 25–49. Geneva: CERN.

Dzuba, V. A. et al. 1985. "Relativistic many-body calculations in atoms and parity violation in caesium," *Journal of Physics B* 18: 597–613.

Ellis, C. D., and Henderson, W. J. 1934. "Artificial Radioactivity," *Proc. Roy. Soc.* A146: 206–16.

Emmons, T. P.; Reeves, J. M.; and Fortson, E. N. 1984. "Parity-Nonconserving Optical Rotation in Atomic Lead," *PRL* 51: 2089–92.

Epstein, S. T.; Finkelstein, R. J.; and Oppenheimer, J. R. 1948. "Note on the Stimulated Decay of Negative Mesons," *PR* 74: 1140–1.

Everett, A. E. 1965. "Evidence for the Existence of Shadow Pions in K$^+$ Decay," *PRL* 14: 615–16.

Fazzini, T. et al. 1958. "Electron Decay of the Pion," *PRL* 1: 247–9.

Feldman, L., and Wu, C. S. 1950. "Beta-Spectrum of Be^{10}," *PR* 78: 318.

Fermi, E. 1934a. "Attempt at a Theory of β-Rays," *NC* 11: 1–21.

1934b. "Versuch einer Theorie der β-Strahlen, I," *Z. Phys.* 88: 161–77.

1940. "The Ionization Loss of Energy in Gases and in Condensed Matter," *PR* 57: 485–93.

Fermi, E.; Teller, E.; and Weisskopf, V. 1947. "The Decay of Negative Mesotrons in Matter," *PR* 71: 314–15.

Feynman, R. P. 1985. *"Surely You're Joking, Mr. Feynman."* New York: Norton.

Feynman, R. P., and Gell-Mann, M. 1958. "Theory of the Fermi Interaction," *PR* 109: 193–8.

Fierz, M. 1937. "Zur Fermischen Theorie des β-Zerfalls," *Z. Phys.* 104: 553–65.

Finkelstein, R., and Kaus, P. 1953. "Note on the Beta Interaction," *PR* 92: 1316–19.

Fitch, V. et al. 1965. "Evidence for Constructive Interference between Coherently Regenerated and *CP*-nonconserving Amplitudes," *PRL* 15: 73–6.

Fitch, V.; Quarles, C. A.; and Wilkins, H. C. 1965. "Study of the K$^+$ Decay Probability," *PR* 140: B1088–91.

Flammersfeld, A. 1939. "Die untere Grenze Des Kontinuerlichen β-Spektrums des RaE," *Z. Phys.* 112: 727–43.

Fletcher, J. C., and Forster, H. K. 1949. "Energy of the Disintegration Product of the Light Mesotron," *PR* 75: 204–5.

Fortson, E. N., and Lewis, L. L. 1984. "Atomic Parity Nonconservation Experiments," *Physics Reports* 113: 289–344.

Fowler, E. C.; Cool, R. L.; and Street, J. C. 1948. "An Example of the Beta-Decay of the Light Meson," *PR* 74: 101–2.

Frankel, S. G. et al. 1958. "Detection of Ga^{66} Polarization by the Annihilation-in-Flight Rate in Polarized Matter," in Lipkin (1958), 385–7.

Franklin, A. 1986. *The Neglect of Experiment.* Cambridge: Cambridge University Press.

Franklin, A. et al. 1989. "Can a Theory-Laden Observation Test the Theory?" *British Journal for the Philosophy of Science* 40: 229–31.

Franklin, A., and Howson, C. 1984. "Why Do Scientists Prefer to Vary Their Experiments?" *Studies in History and Philosophy of Science* 15: 51–62.

208 References

1985. "Newton and Kepler, A Bayesian Approach," *Studies in History and Philosophy of Science* 16: 379–85.

Frauenfelder, H. 1958. "Measurement of Electron Polarization," in Lipkin (1958), 378–9.

Frauenfelder, H. et al. 1957a. "Parity and Polarization of Electrons from Co^{60}," *PR* 106: 386–7.

1957b. "Parity and Electron Polarization: Moller Scattering," *PR* 107: 643–4.

1957c. "Parity and Electron Polarization: Au^{198}," *PR* 107: 909–10.

1957d. "Parity and Electron Polarization: $0 \rightarrow 0$ Transition in Ga^{66}," *PR* 107: 910–11.

Frauenfelder, H., and Henley, E. M. 1975. *Nuclear and Particle Physics.* Reading, MA: Benjamin.

Friedman, H. L., and Rainwater, J. 1951. "Experimental Search for the Beta-Decay of the π^+ Meson," *PR* 84: 684–90.

Friedman, J. L., and Telegdi, V. L. 1957a. "Nuclear Emulsion Evidence for Parity Nonconservation in the Decay Chain $\pi^+ - \mu^+ - e^+$," *PR* 105: 1681–2.

1957b. "Nuclear Emulsion Evidence for Parity Nonconservation in the Decay Chain $\pi^+ - \mu^+ - e^+$," *PR* 106: 1290–3.

Fulbright, H. W., and Milton, J. C. D. 1949. "The Beta-Spectrum of Be^{10}," *PR* 76: 1271–2.

Gaillard, J. M. et al. 1967. "Measurement of the Decay of the Long-Lived Neutral K Meson into Two Neutral Pions," *PRL* 18: 20–5.

Galbraith, W. et al. 1965. "Two-pion Decay of the K°_2," *PRL* 14: 383–6.

Galison, P. 1987. *How Experiments End.* Chicago: University of Chicago Press.

Gamow, G., and Teller, E. 1936. "Selection Rules for the β-Disintegration," *PR* 49: 895–9.

Garwin, R. L. 1957. "Columbia Experiments on Polarized Muons," *Bulletin of the American Physical Society* 2: 206.

Garwin, R. L.; Lederman, L. M.; and Weinrich, M. 1957. "Observation of the Failure of Conservation of Parity and Charge Conjugation in Meson Decays: The Magnetic Moment of the Free Muon," *PR* 105: 1415–17.

Gilbert, S. L. et al. 1985. "Measurement of Parity Nonconservation in Atomic Cesium," *PRL* 55: 2680–3.

Gilbert, S. L., and Wieman, C. E. 1986. "Atomic Beam Measurement of Parity Nonconservation in Cesium," *PR* 34A: 792–803.

Goldhaber, M.; Grodzins, L.; and Sunyar, A. 1957. "Evidence for Circular Polarization of Bremsstrahlung Produced by Beta Rays," *PR* 106: 826–8.

1958. "Helicity of Neutrinos," *PR* 109: 1015–17.

Good, I. J. 1967. "On the Principle of Total Evidence," *British Journal for the Philosophy of Science* 17: 319–21.

Good, M. L. 1951. "Beta-Ray Spectrum of K^{40}," *PR* 83: 1054–5.

Good, M. L., and Lauer, E. J. 1957. "Electron–Neutrino Angular Correlation in the Beta Decay of Neon-19," *PR* 105: 213–16.

Gooding, D.; Pinch, T.; and Schaffer, S. (eds.). 1988. *The Uses of Experiment:*

Studies of Experimentation in the Natural Sciences. Cambridge: Cambridge University Press.

Gross, L., and Hamilton, D. R. 1950. "Beta-Spectrum of S^{35}," *PR* 78: 318.

Hacking, I. 1983. *Representing and Intervening.* Cambridge: Cambridge University Press.

Hamilton, D. R. 1947. "Electron–Neutrino Angular Correlation in Beta-Decay," *PR* 71: 456–7.

Hanna, S. S., and Preston, R. S. 1957. "Positron Polarization Demonstrated by Annihilation in Magnetized Iron," *PR* 106: 1363–4.

Harding, S. (ed.). 1976. *Can Theories Be Refuted?* Dordrecht: Reidel.

Haxby, R. O. et al. 1940. "Threshold for the Proton–Neutron Reactions of Lithium, Beryllium, Boron, and Carbon," *PR* 58: 1035–42.

Henley, E. M., and Wilets, L. 1976. "Parity Nonconservation in T1 and Bi Atoms," *PR* 14A: 1411–17.

Hermannsfeldt, W. B. et al. 1957. "Electron–Neutrino Angular Correlation in the Positron Decay of Argon 35," *PR* 107: 641–3.

   1958. "Determination of the Gamow-Teller Beta-Decay Interaction from the Decay of Helium-6," *PRL* 1: 61–3.

   1959. "Electron–Neutrino Angular Correlations in the Beta-Decays of He^6, Ne^{19}, Ne^{23}, and A^{35}," *Bulletin of the American Physical Society* 4: 77–8.

Hincks, E. P., and Pontecorvo, B. 1948. "Search for Gamma-Radiation in the 2.2 Microsecond Meson Decay Process," *PR* 73: 257–8.

   1949. "The Penetration of μ-Meson Decay Electrons and Their Bremsstrahlung Radiation," *PR* 75: 698–9.

   1950. "On the Disintegration Products of the 2.2-μSec. Meson," *PR* 77: 102–19.

Hollister, J. H. et al. 1981. "Measurement of Parity Nonconservation in Atomic Bismuth," *PRL* 46: 643–6.

Hones, M. J. 1987. "The Neutral-weak-current Experiments: A Philosophical Perspective," *Studies in History and Philosophy of Science* 18: 221–51.

Hornyak, W. F., and Lauritsen, T. 1950. "The Beta-Decay of B^{12} and Li^8," *PR* 77: 160–4.

Horowitz, J. J.; Kofoed-Hansen, O.; and Lindhard, J. 1948. "On the β-Decay of Mesons," *PR* 74: 713–17.

Horwich, P. 1982. *Probability and Evidence.* Cambridge: Cambridge University Press.

Howson, C. 1985. "Some Recent Objections to the Bayesian Theory of Support," *British Journal for the Philosophy of Science* 36: 305–9.

   1987. "Fitting Your Theory to the Facts: Probably not such a Bad Thing After All," forthcoming.

Howson, C., and Franklin, A. 1985. "A Bayesian Analysis of Excess Content and the Localisation of Support," *British Journal for the Philosophy of Science* 36: 425–31.

Howson, C., and Urbach, P. 1989. *Scientific Reasoning: The Bayesian Approach.* Peru, IL: Open Court.

Hubbard, H. W. 1952. University of California Radiation Laboratory Report UCRL-1623, unpublished.

Hughes, D. J. 1940. "Positive Excess and Electron Component in the Cosmic-Ray Spectrum," *PR* 57: 592–7.

Impeduglia, G. et al. 1958. "β Decay of the Pion," *PRL* 1: 249–51.

Iwanenko, D. 1934. "Interactions of Neutrons and Protons," *Nature* 133: 981–2.

Jackson, J. D.; Treiman, S. B.; and Wyld, H. H. 1957. "Possible Tests of Time Reversal Invariance in Beta Decay," *PR* 106: 517–21.

Johnson, W. R. et al. 1985. "Weak-Interaction Effects in Heavy Atomic Systems," *PR* 32A: 2093–9.

1986. "Weak-Interaction Effects in Heavy Atomic Systems. II," *PR* 34A: 1043–57.

Jones, H. 1939. "Energy Distribution and Positive Excess of Mesotrons," *RMP* 11: 235–8.

Kabir, P. K. 1963. *The Development of Weak Interaction Theory*. New York: Gordon and Breach.

1968. *The CP Puzzle*. New York: Academic Press.

Kalitzin, N. 1965. "The *CP* Invariance, the Paritino, and the Spurion," *Academie Bulgare des Sciences, Comptes Rendus* 18: 627–9.

Kautsky, K. 1902. *The Social Revolution*. Chicago: Charles H. Kerr and Company.

Khalfin, L. A. 1965. "On the Quantum Theory of Unstable Elementary Particles," *Doklady Akademii Nauk SSSR* 162: 1034–7.

Kikuchi, S. et al. 1939. "Beta-Ray Spectrum of N^{13}," *Proceedings of the Physico-Mathematical Society of Japan* 21: 52–8.

Klein, O. 1948. "Mesons and Nucleons," *Nature* 161: 897–9.

Koertge, N. 1978. "Towards a New Theory of Scientific Inquiry," in *Progress and Rationality in Science*, eds. G. Radnitzky and G. Andersson, 253–78. Dordrecht: Reidel.

Kofoed-Hansen, O. 1955. "Neutrino Recoil Experiments," in Siegbahn (1955), 357–72.

Konopinski, E. 1943. "Beta-Decay," *RMP* 15: 209–45.

Konopinski, E. J. 1958. "Theory of the Classical β-Decay Measurements," in Lipkin (1958), 319–35.

1959. "Experimental Clarification of the Laws of β-Radioactivity," *Annual Reviews of Nuclear Science* 9: 99–158.

Konopinski, E. J., and Langer, L. M. 1953. "The Experimental Clarification of the Theory of β-Decay," *Annual Review of Nuclear Science* 2: 261–304.

Konopinski, E. J., and Mahmoud, H. M. 1953. "The Universal Fermi Interaction," *PR* 92: 1045–57.

Konopinski, E., and Uhlenbeck, G. E. 1935. "On the Fermi Theory of Radioactivity," *PR* 48: 7–12.

1941. "On the Fermi Theory of β-Radioactivity," *PR* 60: 308–20.

Kreisler, M. 1975. "Faster Than Light Particles – Do They Exist?" *The Physics Teacher* 13: 429–34.

Krohn, V. E. 1957. "Nonconservation of Parity in the Decay of the Neutron," *Bulletin of the American Physical Society* 2: 340.

Kuipers, B.; Moskowitz, A. J.; and Kassirer, J. P. 1988. "Critical Decisions under Uncertainty: Representation and Structure," *Cognitive Science* 12: 177–210.

Kunze, P. 1933. "Untersuchung der Ultrastrahlung in den Wilsonkammer," *Z. Phys* 83: 1–18.

Kurie, F. N. D.; Richardson, J. R.; and Paxton, H. C. 1936. "The Radiations from Artificially Produced Radioactive Substances," *PR* 49: 368–81.

Lagarrigue, A. 1952. "Etude éxperimentale du Spectre de l'Energie de l'Electron de Désintegration du Méson μ," *CR* 234: 2060–1.

Lagarrigue, A., and Peyrou, C. 1951. "Détermination Expérimentale du Spectre d'Energie de l'Electron de Désintégration du Méson μ," *Le Journal de Physique et le Radium* 12: 848–51.

Landau, L. 1957. "On the Conservation Laws for Weak Interactions," *Nuclear Physics* 3: 127–31.

Langer, L. M.; Moffat, R. D.; and Price, H. C. 1949. "The Beta-Spectra of Cu^{64}," *PR* 76: 1725–6.

Langer, L. M.; Motz, J. W.; and Price, H. C. 1950. "Low Energy Beta-Ray Spectra: Pm^{147} S^{35}," *PR* 77: 798–805.

Langer, L. M., and Price, H. C. 1949. "Shape of the Beta-Spectrum of the Forbidden Transition of Yttrium 91," *PR* 75: 1109.

Langer, L. M., and Whitaker, M. D. 1937. "Shape of the Beta-Ray Distribution Curve of Radium E at High Energies," *PR* 51: 713–17.

Langley, P. et al. 1987. *Scientific Discovery*. Cambridge: MIT Press.

Lattes, C. M. G. et al. 1947. "Processes Involving Charged Mesons," *Nature* 159: 694–7.

Lawson, J. L. 1939. "The Beta-Ray Spectra of Phosphorus, Sodium, and Cobalt," *PR* 56: 131–6.

Lawson, J. L., and Cork, J. M. 1940. "The Radioactive Isotopes of Indium," *PR* 57: 982–94.

Lederman, L. M. (1957). in Ascoli et al. (1958), VII-31.

Lee, T. D. 1957. "Introductory Survey, Weak Interactions," in Ascoli et al. (1958), VII-1–VII-12.

⸻ 1971. "History of Weak Interactions," a lecture given at Columbia University, March 26, 1971 (unpublished).

Lee, T. D.; Rosenbluth, M.; and Yang, C. N. 1949. "Interaction of Mesons with Nucleons and Light Particles," *PR* 76: 905.

Lee, T. D., and Yang, C. N. 1956. "Question of Parity Nonconservation in Weak Interactions," *PR* 104: 254–8.

⸻ 1957. "Parity Nonconservation and a Two-Component Theory of the Neutrino," *PR* 105: 1671–5.

Leighton, R. B.; Anderson, C. D.; and Seriff, A. J. 1949. "The Energy Spectrum of the Decay Particles and the Mass and Spin of the Mesotron," *PR* 75: 1432–7.

Leipunski, A. I. 1936. "Determination of the Energy Distribution of Recoil Atoms During β-Decay and the Existence of the Neutrino," *Proceedings of the Cambridge Philosophical Society* 32: 301–3.

Levi Setti, R., and Tomasini, G. 1951. "On the Decay of μ-Mesons," *NC* 8: 994–1005.

Levy, M., and Nauenberg, M. 1964. "Apparent *CP* Violation Due to a New Vector Boson," *PL* 12: 155–6.

Lewis, H., and Bohm, D. 1946. "The Low Energy β-Spectrum of Cu^{64}," *PR* 69: 129–30.

Lewis, L. L. et al. 1977. "Upper Limit on Parity-Nonconserving Optical Rotation in Atomic Bismuth," *PRL* 39: 795–8.

Lipkin, H. (ed.). 1958. *Proceedings of the Rehovoth Conference on Nuclear Structure.* Amsterdam: North Holland.

Lipkin, H. S. et al. 1958. Measurement of Beta-Ray Polarization by Double Coulomb Scattering," in Lipkin (1958), 400–3.

Livingston, M. S., and Bethe, H. A. 1937. "Nuclear Physics," *RMP* 9: 245–390.

Lokanathan, S., and Steinberger, J. 1955. "Search for the β-Decay of the Pion," *NC* 10: 151–62.

Lokanathan, S.; Steinberger, J.; and Wolfe, H. B. 1954. "High-Energy Electrons in the Capture of μ-Mesons by Complex Nuclei," *PR* 95: 624.

Longmire, C., and Brown, H. 1949. "Screening and Relativistic Effects on Beta-Spectra," *PR* 75: 264–70.

Lyman, E. M. 1937. "The Beta-Ray Spectrum of Radium E and Radioactive Phosphorus," *PR* 51: 1–7.

   1939. "The β-Ray Spectrum of N^{13} and the Mass of the Neutrino," *PR* 55: 234.

Lyuboshitz, L.; Okonov, E. O.; and Podgoretskii, M. I. 1965. "Galactic Hypercharge Field and the Two-Pion Decay of the Longlived Neutral K Mesons," *Journal of Nuclear Physics* 1: 490–6.

MacKenzie, D. 1989. "From Kwajelein to Armageddon? Testing and the Social Construction of Missile Accuracy," in *The Uses of Experiment,* eds. D. Gooding, T. Pinch, and S. Shaffer. Cambridge: Cambridge University Press, 409–35.

Macq, P. C.; Crowe, K. M.; and Haddock, R. P. 1958. "Helicity of the Electron and Positron in Muon Decay," *PR* 112: 2061–71.

Madgwick, E. 1927. "The Absorption and Reduction of Velocity of β-Rays on Their Passage Through Matter," *Proceedings of the Cambridge Philosophical Society* 23: 970–81.

Maher, P. 1988. "Why Scientists Gather Evidence," *British Journal for the Philosophy of Science,* forthcoming.

Mahmoud, H. M., and Konopinski, E. J. 1952. "The Evidence of the Once-Forbidden Spectra for the Law of β-Decay," *PR* 88: 1266–75.

Maier-Leibnitz, H. 1939. "Untersuchungen mit der 'langsamen' Wilson Kammer," *Z. Phys.* 112: 569–86.

Marshak, R. E. 1942. "Forbidden Transitions in β-decay and Orbital Electron Capture and Spins of Nuclei," *PR* 61: 431–49.

   1952. *Meson Physics.* New York: Dover.

Marshak, R. E., and Bethe, H. A. 1947. "On the Two-Meson Hypothesis," *PR* 72: 506–9.

Martensson, A. M.; Henley, E. M.; and Wilets, L. 1981. "Calculation of Parity-Nonconserving Optical Rotation in Atomic Bismuth," *PR* 24A: 308–17.

Maxson, D. R.; Allen, J. S.; and Jentschke, W. K. 1955. "Electron–Neutrino Angular Correlation in the Beta Decay of Neon[19]," *PR* 97: 109–16.

Mayer, M. G.; Moszkowski, S. A.; and Nordheim, L. W. 1951. "Nuclear Shell Structure and Beta Decay. I. Odd *A* Nuclei," *RMP* 23: 315–21.

Mayer, M. G., and Telegdi, V. L. 1957. " 'Twin' Neutrinos: A Modified Two-Component Theory," *PR* 107: 1445–7.

Michel, L. 1950. "Interaction between Four Half-Spin Particles and the Decay of the μ-Meson," *Proceedings of the Physical Society (London)* A63: 514–23.

　　1952. "Coupling Properties of Nucleons, Mesons, and Leptons," *Progress in Cosmic Ray Physics* 1: 127–90.

　　1957. "Weak Interactions Between 'Old' Particles and Beta Decay," *RMP* 29: 223–30.

Michel, L., and Wightman, A. 1954. "μ-Meson Decay, β Radioactivity, and Universal Fermi Interaction," *PR* 93: 354–5.

Miller, D., and Popper, K. R. 1983. "A Proof of the Impossibility of Inductive Probability," *Nature* 302: 687–8.

Miller, D. J. 1977. "Elementary Particles – A Rich Harvest," *Nature* 269: 286–8.

Millikan, R. A. 1916. "The Existence of a Subelectron?" *PR* 8: 595–625.

Moller, C. 1937. "Einige Bemerkung zur Fermischen Theorie des Positronen-zerfalls," *Physikalische Zeitschrift der Sowjetunion* 11: 9–17.

Morpurgo, G. 1957. "Possible Explanation of the Decay Processes of the Pion in the Frame of the 'Universal' Fermi Interaction," *NC* 5: 1159–65.

Morrison, M. 1986. "More on the Relationship between Technically Good and Conceptually Important Experiments," *British Journal for the Philosophy of Science* 37: 101–15.

Mott, N. F. 1929. "Scattering of Fast Electrons by Atomic Nuclei," *Proc. Roy. Soc.* A124: 425–42.

　　1932. "The Polarisation of Electrons by Double Scattering," *Proc. Roy. Soc.* A135: 429–58.

Neary, G. J. 1940. "The β-Ray Spectrum of RaE," *Proc. Roy. Soc.* A175: 71–87.

Neddermeyer, S. H., and Anderson, C. D. 1938. "Cosmic Ray Particles of Intermediate Mass," *PR* 54: 88–9.

Neilsen, W. M. et al. 1941. "Differential Measurement of the Meson Lifetime," *PR* 59: 547–53.

Nereson, N., and Rossi, B. 1943. "Further Measurements on the Disintegration Curve of Mesotrons," *PR* 64: 199–201.

Neuffer, D. V., and Commins, E. D. 1977. "Calculation of Parity-Nonconserving Effects in the $6^2P_{1/2}$–$7^2P_{1/2}$ Forbidden *M*1 Transition in Thallium," *PR* 16A: 844–62.

Niiniluoto, I. 1983. "Novel Facts and Bayesianism," *British Journal for the Philosophy of Science* 34: 375–9.

Nishijima, K., and Saffouri, M. J. 1965. "*CP* Invariance and the Shadow Universe," *PRL* 14: 205–7.

Nishina, Y.; Takeuchi, M.; and Ichimiya, Y. 1939. "On the Mass of the Mesotron," *PR* 55: 585–6.

Nordheim, L. W., and Hebb, M. H. 1939. "On the Production of the Hard Component of the Cosmic Radiation," *PR* 56: 494–501.

Nordsieck, A. 1934. "Neutron Collisions and the Beta Ray Theory of Fermi," *PR* 46: 234–5.

Novikov, V. N.; Sushkov, O. P.; and Khriplovich, I. B. 1976. "Optical Activity of Heavy-Metal Vapors – A Manifestation of the Weak Interaction of Electrons and Nucleons," *JETP* 44: 872–80.

Nye, M. J. 1980. "N-Rays: An Episode in the History and Psychology of Science," *Historical Studies in the Physical Sciences* 11: 125–56.

O'Ceallaigh, C. 1950. "The Nature of the Particle Emitted in the Decay of the π-Meson," *Phil. Mag.* 41: 838–48.

O'Conor, J. S. 1937. "The Beta-Ray Spectrum of Radium E," *PR* 52: 303–14.

Owen, G. E., and Cook, C. S. 1949a. "On the Shape of the Positron Spectrum of Cu^{61}," *PR* 76: 1536–7.

    1949b. "The Positron and M(N)egatron Spectra of Cu^{64}," *PR* 76: 1726–7.

Owen, G. E.; Cook, C. S.; and Owen, P. H. 1950. "The Disintegration of Cu^{61}," *PR* 78: 686–90.

Owen, G. E., and Primakoff, H. 1948. "Relation between Apparent Shapes of Monoenergetic Conversion Lines and of Continuous β-Spectra in a Magnetic Spectrometer," *PR* 74: 1406–12.

Page, L. A., and Heinberg, M. 1957. "Measurement of the Longitudinal Polarization of Positrons Emitted by Sodium-22," *PR* 106: 1220–4.

Particle Data Group. 1980. "Review of Particle Properties," *RMP* 52: S1–286.

    1986. "Review of Particle Properties," *PL* 170B: 1–350.

Pauli, W. 1933. "Die Allgemeinen Prinzipen der Wellenmechanik," *Handbuch der Physik* 24: 83–272.

    1957. "On the Conservation of the Lepton Charge," *NC* 6: 204–15.

Paxton, H. C. 1937. "The Radiations Emitted from Artificially Produced Radioactive Substances. III. Details of the Beta-Ray Spectrum of P^{32}," *PR* 51: 170–7.

Peacock, C. L., and Mitchell, A. C. G. 1949. "Disintegration of Cs^{137}," *PR* 75: 1272–4.

Peaslee, D. C. 1953. "The Linear Combination in β Decay," *PR* 91: 1447–57.

    1955. "Kombination der Wechselwerkungsterme in der Theorie des β-Zerfalls," *Z. Phys.* 141: 399–402.

Perez-Mendez, V., and Brown, H. 1950. "The Beta-Spectrum of He^{6}," *PR* 77: 404–5.

Petschek, A. G., and Marshak, R. E. 1952. "The β-Decay of Radium E and the Pseudoscalar Interaction," *PR* 85: 698–9.

Piccioni, O. 1948. "Search for Photons from Meson Capture," *PR* 74: 1754–8.

Pickering, A. 1984a. *Constructing Quarks.* Chicago: University of Chicago Press.

    1984b. "Against Putting the Phenomena First: The Discovery of the Weak Neutral Current," *Studies in History and Philosophy of Science* 15: 85–117.

    1987. "Against Correspondence: A Constructivist View of Experiment and the Real," *Philosophy of Science Association 1986, Volume 2*, eds. A. Fine and P. Machamer. East Lansing, MI: Philosophy of Science Association.

Pinch, T. 1986. *Confronting Nature*. Dordrecht: Reidel.

Pirsig, R. M. 1974. *Zen and the Art of Motorcycle Maintenance: An Inquiry into Values*. New York: William Morrow.

Pontecorvo, B. 1947. "Nuclear Capture of Mesons and the Meson Decay," *PR* 72: 246–7.

Postma, H. et al. 1958. "β^+ Asymmetries for Polarized Co^{60} and Mn^{52} Nuclei," in Lipkin (1958), 408–9.

Prentki, J. 1966. "*CP* Violation," *Proceedings of Oxford International Conference on Elementary Particles, 1965*, 47–58. Oxford: Oxford University Press.

Prescott, C. Y. et al. 1978. "Parity Nonconservation in Inelastic Electron Scattering," *PL* 77B: 347–52.

1979. "Further Measurements of Parity Nonconservation in Inelastic Electron Scattering," *PL* 84B: 524–8.

Pryce, M. H. L. 1952. "Spinor Formulation of Beta-Decay and Similar Interactions," *Z. Phys.* 133: 309–18.

Puppi, G. 1948. "Sui Mesoni dei Raggei Cosmici," *NC* 5: 587–8.

Pursey, D. L. 1951. "The Interaction in the Theory of Beta Decay," *Phil. Mag.* 42: 1193–208.

1952. "Symmetry Properties of a Universal Interaction," *Physica* 18: 1017–19.

Ramsey, N. 1956. *Molecular Beams*. Oxford: Oxford University Press.

Randall, H. M. et al. 1949. *Infrared Determination of Organic Structures*. New York: Van Nostrand.

Rasetti, F. 1941. "Disintegration of Slow Mesotrons," *PR* 60: 198–204.

Reitz, J. F. 1950. "The Effect of Screening on Beta-Ray Spectra and Internal Conversion," *PR* 77: 10–18.

Richardson, O. W. 1934. "The Low Energy β-Rays of Radium E," *Proc. Roy. Soc.* A147: 442–54.

Ridley, B. W. 1954. "Nuclear Recoil in Beta Decay," PhD thesis, Cambridge University (unpublished).

1956. "The Neutrino," *Progress in Nuclear Physics* 5: 188–251.

Rinaudo, G. et al. 1965. "Existence of Pions with Spin," *PRL* 14: 761–3.

Robinette, R. W., and Rosner, J. L. 1982. "Prospects for a Second Neutral Vector Boson at Low Mass in $SO(10)$," *PR* 25D: 3036–64.

Robson, J. M. 1951. "The Radioactive Decay of the Neutron," *PR* 83: 349–58.

1955. "Angular Correlation in the Beta Decay of the Neutron," *PR* 100: 933–5.

Roe, B. P. et al. 1961. "New Determination of the K^+-Decay Branching Ratios," *PRL* 7: 346–8.

Rose, M. E. 1936. "A Note on the Possible Effect of Screening in the Theory of Beta-Disintegration," *PR* 49: 727–9.

Rose, M. E., and Holmes, D. K. 1951. "The Effect of Finite Nuclear Size in Beta Decay," *PR* 83: 190–1.

Rose, M. E., and Osborne, R. K. 1954. "The Pseudoscalar Interaction and the Beta-Spectrum of RaE," *PR* 93: 1315–25.

Rose, M. E., and Perry, C. L. 1953. "The Effect of the Finite de Broglie Wavelength in the Theory of Beta-Decay," *PR* 90: 479–82.

Rose, M. E.; Perry, C. L.; and Dismuke, N. M. 1953. Oak Ridge National Laboratory Report 1459.

Rossi, B. 1939. "The Disintegration of the Mesotron," *RMP* 11: 296–303.

Rossi, B., and Hall, D. B. 1941. "Variation of the Rate of Decay of Mesotrons with Momentum," *PR* 59: 223–8.

Rossi, B., and Nereson, N. 1942. "Experimental Determination of the Disintegration Curve of Mesotrons," *PR* 62: 417–22.

Ruderman, M., and Finkelstein, R. 1949. "Note of the Decay of the π-Meson," *PR* 76: 1458–60.

Rustad, B. M., and Ruby, S. L. 1953. "Correlation between Electron and Recoil Nucleus in He^6 Decay," *PR* 89: 880–1.

1955. "Gamow-Teller Interaction in the Decay of He^6," *PR* 97: 991–1002.

Sagane, R.; Dudziak, W. F.; and Vedder, J. 1954. "Positron Spectrum from the Decay of the μ Meson," *PR* 95: 863–4.

1955. "Positron Spectrum from the Decay of the μ Meson," *PR* 98: 269.

Sagane, R.; Gardner, W. L.; and Hubbard, H. W. 1951. "Energy Spectrum of the Electrons from μ^+ Meson Decay," *PR* 82: 557–8.

Sagane, R. et al. 1956. "μ^+–β^+ Decay Spectrum," *Bulletin of the American Physical Society* 1: 174.

Sakata, S., and Inoue, T. 1946. "On the Correlations between Mesons and Yukawa Particles," *Progress in Theoretical Physics* 1: 143–50.

Salam, A. 1957. "On Parity Conservation and the Neutrino Mass," *NC* 5: 299–301.

Sandars, P. G. H. 1980. "Many Body Aspects of Parity Nonconservation in Heavy Atoms," *Physica Scripta* 21: 284–92.

Sard, R. D., and Althaus, E. J. 1948. "A Search for Delayed Photons from Stopped Sea Level Cosmic-Ray Mesons," *PR* 74: 1364–71.

Sargent, B. W. 1932. "Energy distribution curves of the disintegration electrons," *Proceedings of the Cambridge Philosophical Society* 24: 538–53.

1933. "The Maximum Energy of the β-Rays from Uranium X and other Bodies," *Proc. Roy. Soc.* A139: 659–73.

Sargent, C. P. et al. 1955. "Diffusion Cloud-Chamber Study of Very Slow Mesons. II. Beta Decay of the Muon," *PR* 99: 885–8.

Savage, L. 1982. *The Foundations of Statistics*, 2nd ed. New York: Dover.

Schopper, H. 1957. "Circular Polarization of γ-rays: Further Proof for Parity Failure in β Decay," *Phil. Mag.* 2: 710–13.

1958. "β–γ Circular Polarization Correlation," in Lipkin (1958), 410–13.

Schwinger, J. 1957. "A Theory of Fundamental Interactions," *Annals of Physics* 2: 407–34.

Shaklee, F. S. et al. 1964. "Branching Ratios of the K^+ Meson," *PR* 136: B1423–31.

Shapere, D. 1982. "The Concept of Observation in Science and Philosophy," *Philosophy of Science* 49: 485–525.

Sherr, R., and Gerhart, J. 1952. "Gamma-Radiation of C^{10}," *PR* 86: 619.

Sherr, R., and Gerhart, J. B. 1956. *Bulletin of the American Physical Society* 1: 219, cited in Konopinski (1958).

Sherr, R., and Miller, R. H. 1954. "Electron Capture in the Decay of Na^{22}," *PR* 93: 1076–81.

Sherr, R.; Muether, H. R.; and White, M. G. 1949. "Radioactivity of C^{10} and O^{14}," *PR* 75: 282–92.

Sherwin, C. W. 1948a. "Momentum Conservation in the Beta-Decay of P^{32} and the Angular Correlation of Neutrinos with Electrons," *PR* 73: 216–25.

1948b. "The Conservation of Momentum in the Beta-Decay of Y^{90}," *PR* 73: 1173–7.

1949. "Neutrinos from P^{32}," *PR* 75: 1799–810.

1951. "Experiments on the Emission of Neutrinos from P^{32}," *PR* 82: 52–7.

Siegbahn, K. 1946. "The Disintegration of Na^{24} and P^{32}," *PR* 70: 127–32.

(ed.). 1955. *Beta- and Gamma-Ray Spectroscopy*. New York: Interscience.

Sigurgeirsson, T., and Yamakawa, A. 1947. "Decay of Mesons Stopped in Light Materials," *PR* 71: 319–20.

Smith, A. M. 1951. "Forbidden Beta-Ray Spectra," *PR* 82: 955–6.

Smith, F. M. 1951. "On the Branching Ratio of the π^+ Meson," *PR* 81: 897–8.

Soreide, D. C. et al. 1976. "Search for Parity Nonconservation in Atomic Bismuth," *PRL* 36: 352–5.

Stech, B., and Jensen, J. H. D. 1955. "Die Kopplungskonstanten in der Theorie des β-Zerfalls," *Z. Phys.* 141: 175–84.

Steffen, R. M. 1958. "Beta–Gamma Angular Correlation Measurements," in Lipkin (1958), 419–36.

Steinberger, J. 1949a. "On the Range of Electrons in Meson Decay," *PR* 75: 1136–43.

1949b. "On the Use of Subtraction Fields and the Lifetimes of Some Types of Meson Decay," *PR* 76: 1180–6.

Street, J. C., and Stevenson, E. C. 1937. "New Evidence for the Existence of a Particle of Mass Intermediate between the Proton and Electron," *PR* 52: 1003–4.

Stuewer, R. 1975. *The Compton Effect.* New York: Science History Publications.

1985. "Artificial Disintegration and the Cambridge-Vienna Controversy," in *Observation, Experiment, and Hypothesis in Modern Physical Science*, eds. P. Achinstein and O. Hannaway, 239–307. Cambridge, MA: MIT Press.

Sudarshan, E. C. G., and Marshak, R. E. 1957. "The Nature of the Four-Fermion Interaction," in *Proceedings, Padua Conference on Mesons and Recently Discovered Particles*, V-14–V-24; reprinted in Kabir (1963), 118–28.

1958. "Chirality Invariance and the Universal Fermi Interaction," *PR* 109: 1860–2.

1985. "Origins of the Universal *V-A* Theory," Virginia Polytechnic and State University Report VPI-HEP-84/8, unpublished.

Tamm, I. 1934. "Exchange Forces between Neutrons and Protons and Fermi's Theory," *Nature* 133: 981.

Tanner, C. E., and Commins, E. D. 1986. "Measurement of Stark Amplitudes α, β in the $6^2P_{1/2} \rightarrow 7^2P_{1/2}$ Transition in Atomic Thallium," *PRL* 56: 332–5.

Taubes, G. 1986. *Nobel Dreams*. New York: Simon and Schuster.

Taylor, J. C. 1958. "Beta Decay of the Pion," *PR* 110: 1216.

Taylor, J. D. et al. 1987. "Measurement of Parity Nonconserving Optical Rotation in the 648 nm Transition in Atomic Bismuth," preprint.

Taylor, S. et al. 1965. Search for Anomalous π^+ Decay Among τ^+ Decay Secondaries," *PRL* 14: 745–6.

Terent'ev, M. V. 1965. $K^\circ_2 \rightarrow 2\pi$ Decay and Possible *CP*-nonconservation," *Uspekhi Fizicheskikh Nauk* 86: 231–62.

Thompson, R. W. 1948. "Cloud-Chamber Study of Meson Disintegration," *PR* 74: 490–1.

Ticho, H. 1948. "The Capture Probability of Negative Mesotrons," *PR* 74: 1337–47.

Tiomno, J. 1955. "Mass Reversal and the Universal Interaction," *NC* 1: 226–32.

Tiomno, J., and Wheeler, J. 1949a. "Energy Spectrum of Electrons from Meson Decay," *RMP* 21: 144–52.

   1949b. "Charge-Exchange Reaction of the μ-Meson with the Nucleus," *RMP* 21: 153–65.

Tolhoek, H. A., and de Groot, S. R. 1951. "Mixed Invariants in Beta-Decay and Symmetries Imposed on the Interaction Hamiltonian," *PR* 84: 150–1.

Tomonaga, S., and Araki, G. 1940. "Effect of the Nuclear Coulomb Field on the Capture of the Slow Mesons," *PR* 58: 90–1.

Townsend, A. A. 1941. "β-Ray Spectra of Light Elements," *Proc. Roy. Soc.* A177: 357–66.

Treiman, S. B., and Wyld, H. H. 1956. "Decay of the Pi Meson," *PR* 101: 1552–7.

Tyler, A. W. 1939. "The Beta- and Gamma-Radiations from Copper^{64} and Europium^{152}," *PR* 56: 125–30.

Urbach, P. 1981. "On the Utility of Repeating the 'Same' Experiment," *Australasian Journal of Philosophy* 59: 151–62.

Valley, G. E. 1947. "The Radioactive Decay of Slow Positive and Negative Mesons," *PR* 72: 772–84.

Varder, R. W. 1915. "The Absorption of Homogeneous β-Rays," *Phil. Mag.* 29: 725–33.

Vilain, J. H., and Williams, R. W. 1953. "Decay-Electron Spectrum of the μ Meson," *PR* 92: 1586–7.

   1954. "μ-Meson Decay Spectrum," *PR* 94: 1011–17.

Wang, K., and Jones, S. B. 1948. "On the Disintegration of Mesotrons," *PR* 75: 1547–8.

Warwick, J. W. et al. 1981. "Planetary Radio Astronomy Observations from Voyager 1 Near Saturn," *Science* 212: 239–43.

   1982. "Planetary Radio Astronomy Observations from Voyager 2 Near Saturn," *Science* 215: 582–7.

Washburn, S. 1986. *A Moral Alphabet of Vice and Folly*. New York: Arbor House Publishing.

Watase, Y. 1940. "On the Disintegration of the N^{13} Nucleus," *Proceedings of the Physico-Mathematical Society of Japan* 22: 639–46.

Watase, Y., and Itoh, J. 1938. "The β-Ray Spectrum of RaE," *Proceedings of the Physico-Mathematical Society of Japan* 20: 809–13.

Weisskopf, V. 1947. "On the Production Process of Mesons," *PR* 72: 510.

Weinberg, S. 1964. "Do Hyperphotons Exist?" *PRL* 13: 495–7.

Weyl, H. 1929. "Elektron und Gravitation. I," *Z. Phys.* 56: 330–52.

Wheaton, B. 1983. *The Tiger and the Shark*. Cambridge: Cambridge University Press.

Wick, G. C.; Wightman, A. S.; and Wigner, E. P. 1952. "The Intrinsic Parity of Elementary Particles," *PR* 88: 101–5.

Wieman, C. E.; Gilbert, S. L.; and Noecker, M. C. 1987. "A New Measurement of Parity Nonconservation in Atomic Cesium," in *Atomic Physics 10*, eds. H. Narumi and I. Shimamura, 65–76. Amsterdam: North Holland.

Williams, W. L. (ed.). 1980. *Proceedings, International Workshop on Neutral Current Interactions in Atoms*, Cargese, September 10–14, 1979. Washington: National Science Foundation.

Williams, E. J., and Roberts, G. E. 1940. "Evidence for the Transformation of Mesotrons into Electrons," *Nature* 145: 102–3.

Wojcicki, S. G. et al. 1963. "Production of the Kappa Meson in K–p Interactions," *Physics Letters* 5: 283–6.

Woolgar, S. 1981. "Interests and Explanation in the Social Study of Science," *Social Studies of Science* 11: 365–94.

Worrall, J. 1982. "The Pressure of Light: The Strange Case of the Vacillating Crucial Experiment," *Studies in History and Philosophy of Science* 13: 133–71.

Wu, C. S. 1950. "Recent Investigations of β-Ray Spectra," *RMP* 22: 386–98.

1955a. "The Interaction in β-decay," in Bellamy and Moorhouse (1955), 177–85.

1955b. "Experiments on the Shapes of β-Spectra. The Interaction in β-Decay," in Siegbahn (1955), 314–56.

1958. "Measurement of the Longitudinal Polarization of β-Particles by Moller Scattering," in Lipkin (1958), 380–4.

Wu, C. S., and Albert, R. D. 1949a. "The Beta-Spectra of Cu^{64}," *PR* 75: 315–16.

1949b. "The Beta-Ray Spectra of Cu^{64} and the Ratio of N_+/N_-," *PR* 75: 1107–8.

Wu, C. S. et al. 1957. "Experimental Test of Parity Nonconservation in Beta Decay," *PR* 105: 1413–15.

Wu, C. S., and Schwarzschild, A. 1958. "A Critical Examination of the He^6 Recoil Experiment of Rustad and Ruby," Columbia University Report CU-173, unpublished.

Yamada, M. 1953. "Theoretical Reinvestigation of the β-spectrum of RaE," *Progress of Theoretical Physics* 10: 252–64.

Yang, C. N., and Tiomno, J. 1950. "Reflection Properties of Spin-½ Fields and a Universal Fermi-Type Interaction," *PR* 79: 495–8.

Yukawa, H. 1935. "On the Interaction of Elementary Particles, I," *Proceedings of the Physico-Mathematical Society of Japan* 17: 48–57.

Yukawa, H.; Sakata, S.; and Taketani, M. 1938. "On the Interaction of Elementary Particles, III," *Proceedings of the Physico-Mathematical Society of Japan* 20: 319–40.

Zar, J. L.; Hershkowitz, J.; and Berezin, E. 1948. "Cloud-Chamber Study of Electrons from Meson Decay," *PR* 74: 111–12.

Index